Carl Julius Salomonsen, William Trelease

Bacteriological Technology for Physicians

Carl Julius Salomonsen, William Trelease

Bacteriological Technology for Physicians

ISBN/EAN: 9783743686915

Printed in Europe, USA, Canada, Australia, Japan

Cover: Foto ©berggeist007 / pixelio.de

More available books at **www.hansebooks.com**

Bacteriological Technology

For Physicians

With Seventy-Two Figures in the Text

BY

Dr. C. J. Salomonsen

Authorized Translation from the Second Revised Danish Edition

BY

William Trelease

New York
William Wood & Company
1890

AUTHOR'S PREFACE.

THE first edition of this text-book was issued in 1885, as a reprint from the "Nordiskt Medicinskt Arkiv." The increased size of this new edition is partly due to the addition of an introduction to the microscopic investigation of bacteria; but the active development of bacteriology in the last few years has also necessitated a re-elaboration and extension of all of the chapters.

In the work, I have had two objects in view: the preparation of an outline adapted to bacteriological courses for physicians and veterinary surgeons, and a guide for those who are obliged to take up the subject at home without the assistance of an instructor, yet wish to carry out for themselves the fundamental experiments which are most important for pathology and hygiene. I have, therefore, not attempted to give an exhaustive presentation of the entire technology of the subject, but rather to describe the simplest and most easily managed apparatus and methods, so that the equipment of a home laboratory need not appear too expensive; while I have tried to describe these in a sufficiently elementary and detailed manner.

Those who wish a more comprehensive treatise, are referred to the last edition of Hueppe's "Methoden der Bakterienforschung" (Wiesbaden, 1889).

C. J. S.

TRANSLATOR'S PREFACE.

So many works on bacteriology are now accessible, that a new one must possess decided merit to justify its appearance. The scope of the present little volume is such as to satisfy me that it should fill a place as yet vacant; and the careful treatment of the subject gives rise to the hope that it may find a larger field of usefulness in the English language than could be possible while it was confined to the less-used Danish.

The physician who wishes to read a well-written account of the more important pathogenic bacteria, is further referred to Fraenkel's "Bakterienkunde." The fullest and most reliable attempt at a bacteriological flora, is to be found in Fluegge's "Fermente und Mikroparasiten," which contains much additional information.

W. T.

BACTERIOLOGICAL TECHNOLOGY.

CHAPTER I.

STERILIZATION.

BACTERIOLOGICAL technology must in the first place put means in our hands for separating the different kinds of bacteria from one another, and cultivating each species in a state of complete purity. Pure cultivation is a necessary condition for a trustworthy study of the morphology and physiology of these microscopic organisms. Unless one has this constantly in mind during his work, he is certain to end in sad confusion; and the short history of bacteriology is only too full of worthless investigations and wrong results, which are to be attributed to the untrustworthiness of the methods employed.

It is well known that the pure cultivation of a single kind of bacteria is surrounded by quite peculiar difficulties, due to the extreme smallness, the large numbers, and the extraordinary distribution, one might almost say omnipresence, of bacteria. Everything which is allowed to stand uncovered in the air for a short time is likely to receive spores capable of germination, which fall on its surface with the dust. Every contact with hands or clothing is attended with the same danger. We must learn first of all to meet and control this accidental and invisible source of contamination, for it is self-evident that all vessels, instruments, fluids, etc., which are to be used in cultures must be clean, not only in the sense in which chemists use the word, but also in the bacteriological sense, or sterile, that is, free from living germs.

Heat is the most generally used means of sterilization, but

it must be used in different ways, and of different degrees, according to the nature of the material to be sterilized, a platinum needle, for example, is best sterilized by heating it to incandescence directly in the flame; but a knife blade, or a pair of scissors, can clearly not be heated to this point. These can only be "flamed," that is, drawn slowly back and forth through the flame, until one is sure that all of their surface has been heated to at least 200° C., so that any germs have been burned.

But it is only exceptionally that glowing or flaming can be employed. If one has to deal with glass-ware, metallic articles, cotton, paper, etc., these are most frequently sterilized in air heated to about 150° C. In the preparation of most fluid and solid culture-media, one can most rapidly and surely reach the desired end by the help of a Papin digester, an "autoclave," in which the sterilization is effected by steam under pressure, at 110 to 120° C.; or one contents himself with using the boiling temperature, either by directly cooking the substance, or by heating it to 100° C., or about that point, by the aid of streaming steam. Occasionally, however, one may have to use a temperature lower than this (60 to 70° C.), to kill germs, for instance where serum is used. In some cases filtration and disinfecting solutions are used as means of sterilizing.

The entire collection of instruments needed for sterilizing, the apparatus and culture-media to be used in the investigations that we shall describe, are, in addition to a common Bunsen gas burner (or alcohol lamp), the following:

1. *A sterilizing oven* in which articles made of glass and metal, cotton, paper, etc., are heated to about 150° C., by means of hot air.

For this purpose one uses in the laboratory a double-walled sheet-iron box 30 cm. high, 35 cm. broad, and 23 cm. deep, with a tube in the top for inserting a thermometer, a door in the side, and with a loose middle piece in the bottom against which the large gas jet constantly plays, and which consequently will burn through after long use, when it is only necessary to put in a new piece of iron.

One can usually get along with less expense by making a serviceable sterilizing oven (Figs. 1, 2) of one of the common cracker boxes (about 24 cm. high, 22 cm. wide, and 24 cm.

deep) to be obtained cheaply of any grocer. In the cover a round hole is to be cut in which a bored cork is fitted to carry a thermometer, registering to 200° C. For ventilation, a small hole is punched in each of the four side-walls close to the top and bottom. In the box is set a four-sided piece of strong sheet-iron, with the sides bent down so as to hold it about 2 cm. from the bottom, so that the objects to be sterilized will not come into immediate contact with the latter. The box is inclosed in a substance which is a poor conductor of heat, felt being usually chosen. One piece covers the lid, and is perfor-

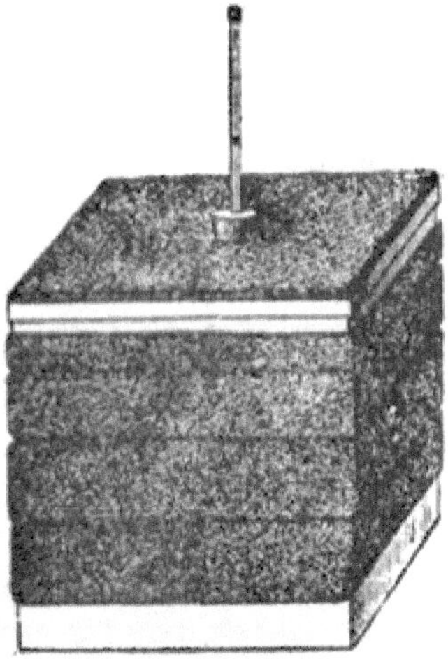

Fig. 1.

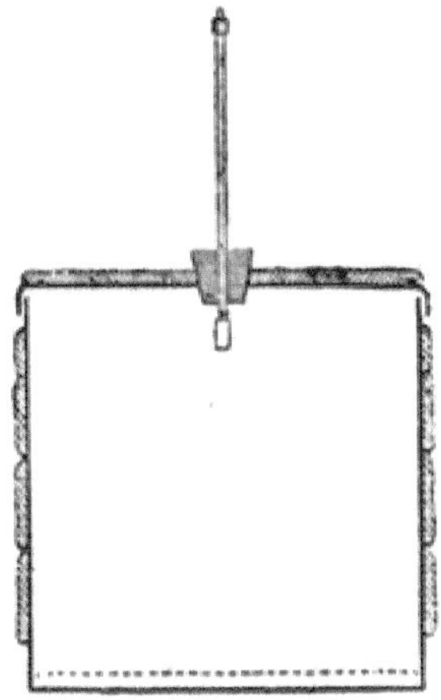

Fig. 2.

Figs. 1 and 2.—Oven for sterilization at 150° C., formed of a cracker box surrounded with felt and provided with thermometer and support of iron netting.

ated for the passage of the cork and thermometer. Another piece is chosen long enough to be wrapped around the four sides of the box and wide enough to cover the upper three quarters of these, excepting a small space where the air-holes occur. The lowermost quarter is best left uncovered to avoid scorching or burning the felt. This strip of felt is fastened simply by tying around it tightly a few pieces of cord or wire. The box is set on one or two of the common iron tripods, such as are used in chemical laboratories, and is heated by one or more gas burners. Between these and the bottom of the box, a loose piece of sheet-tin can be placed to prevent burning through the bottom. If gas is not at hand, a small oil stove with one or two burners can be used.

In comparison with the larger and more expensive ovens provided with ventilators and thermo-regulator, this primitive apparatus offers some inconveniences, but these do not count for much in comparison with its perfect usefulness and extreme cheapness.

Care must be taken to guard the sterilized objects from fresh contamination when they are taken out of the sterilizer. Vessels which are guarded from the penetration of germs, by cotton plugs or otherwise, when it is not especially important to keep the outside free from germs (*e.g.*, flasks and test-tubes) can simply be taken from the oven after they have been heated to 150° C. for half an hour or an hour, and set aside without any other precaution than covering or wrapping them up in paper to prevent them from becoming very dusty. But objects which must be perfectly sterilized, such as watch-glasses, slides, large glass plates, etc., must be carefully wrapped in a couple of thicknesses of thin, firm paper before they are put in the oven. Germs from the air cannot pass through this, and the objects can thus be kept satisfactorily sterile for a long time. It is well to wrap each object in several layers of thin paper, and then to wrap these in larger parcels.

Paper and cotton should be kept as far as possible from the bottom of the oven to prevent them from becoming charred by long heating. Paper and cotton (but not absorbent cotton) assume a yellowish-brown color after being kept at from 140° to 150° C. for a long time, and this discoloration consequently shows that the heat in the sterilizing oven has been sufficient even without the testimony of a thermometer.

2. *The Steam Sterilizer of Koch.*—Sterilization at the boiling temperature can evidently be effected by cooking in water, but since this method is not only inapplicable in many cases, but always more laborious and unreliable than sterilization by streaming steam as introduced by Koch, the latter is now the principal method used when it is desired to sterilize at 100° C.

It will be remembered that two advantages are gained by this: the heat penetrates rapidly into the objects to be sterilized, and steam of this temperature (99° to 100° C. according to altitude) sterilizes much more rapidly than atmospheric air of the same or a much higher temperature, probably for chemical reasons.

The partly diagrammatic illustration shows the Koch steam apparatus of a simple and cheap form. It consists of a tin

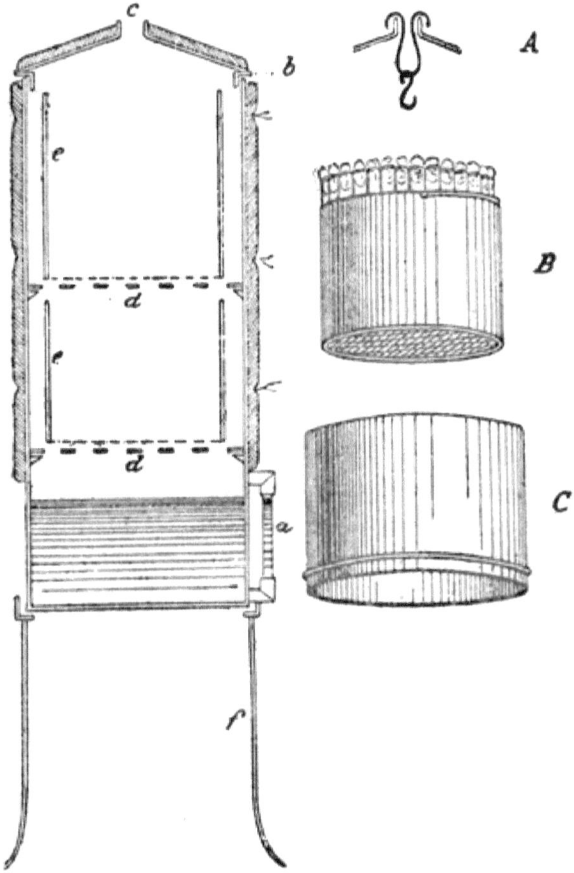

Fig. 3. Fig. 4.

Figs. 3 and 4.—Koch cylinder in its simplest form, with two tin shelves and pails. *A*, Hook for suspending long objects; *B*, tin pail with sieve-bottom, containing plugged test-tubes; *C*, tin cylinder for elongating the apparatus.

cylinder 50 cm. high and 17 cm. in diameter, and is provided with a small glass tube (*a*) fastened by rubber tubing in which the height of the water can be seen from the outside. The absolute size, as well as the proportion between the height and diameter, can naturally be varied within wide limits and the distance between the shelves is to be adapted to the height of the apparatus, etc., that are to be most frequently sterilized. The bottom, which is to be exposed to the action of the flame, is preferably made of somewhat thicker tin than the sides. The top of the apparatus is closed by a conical lid (*b*) the centre of which is pierced by a short tube (*c*) for the reception of a cork and thermometer. The lid and the side walls, to within 9 cm. of the bottom, are usually covered with felt, though this is not absolutely necessary. Within the cylinder, at a distance of 14 and 30 cm. from the bottom, are placed projecting ledges, on which can be placed two loosely fitting shelves of heavy tin perforated with large holes (*dd*). These serve to support tin pails (*ee*) about 12 cm. high and 14 cm. in diameter, the bottom of which is best made of wide-meshed galvanized iron netting (Fig. 4, *B*). These hold the objects to be sterilized, *e.g.*, test-tubes plugged with cotton in the figure. When it is to be used, the apparatus is placed on a tripod (*f*) of stout strap-iron. The pails are taken out and the cylinder is filled with water to a depth of about 10 cm. The pails with their contents having been set in and the lid with its thermometer put in place, the water is heated to the boiling point. After some time steam will escape freely around the loosely-fitting lid, and the thermometer will also show about 100° C., after which the objects are left in this streaming steam for a longer or shorter time (as a rule, twenty minutes), the time being counted from the moment when the thermometer indicates 100° C. (or the steam begins to escape freely). Most objects which are to be sterilized by steam can be set on the tin shelves without difficulty. Longer objects can be hung in the apparatus after the pails and shelves have been removed. This is most easily done by using a piece of stout brass wire, bent and brought into the opening of the lid (Fig. 4, *A*).

The apparatus can be lengthened when this is necessary by using a tin cylinder, with a projecting collar near its base (Fig. 4, *DC*) to be inserted between the lid and the top of the cylinder. A stop-cock, near the bottom, for drawing off the water, increases the expense of the apparatus somewhat, but adds to its convenience. Moreover, one can easily form such a steam sterilizer from a common kitchen pot. It is only necessary to fit to it a tin cylinder open at the bottom, and furnished with a lid and felt covering, but it is essential to provide this cylinder with a projecting collar at a couple of centimetres from the bottom, by which it rests on the edge of the pot; and it is best to provide it with a couple of rings, by which the cylinder may be tied to the ears of the pot.

A word of explanation should be offered for the omission of the Papin digester in speaking of necessary apparatus for sterilization. There are, indeed, bacillus spores which are able to survive a cooking for several hours in water or free steam, while they are killed after a few moments' heating to 120° C. in the digester, consequently this is the only apparatus in which one can rapidly and surely effect sterilization by a single attempt. Still it is a very expensive piece of apparatus, which is the only reason that it has been omitted. Koch's steam cylinder, on the other hand, is very simple, cheap, and efficient when it is used with care. As a rule the spores of bacilli which always get into our culture media, from water, the air, etc., are not so resistant as those which have been mentioned. Generally, streaming steam at 100° C. will kill them, and experiments from a large number of bacteriological laboratories have long since shown the adequacy of Koch's apparatus. However, in the preparation of his culture apparatus one must always have in mind that some very resistant form may have escaped destruction in the steam, and must make sure of absolute freedom from germs in his culture media by:

a. Cleanliness in the preparation. It is not sufficient to trust to the final sterilization of the materials, but at every step of the preparation one should work cleanly, to lessen the chances of resistant germs entering.

b. Discontinuous Heating.—This method was first applied by Tyndall in 1877 for the sterilization of hay infusion, which is known to contain some of the most resistant bacillus spores, the extraordinary vitality of which renders it impossible to sterilize this substance even by cooking it for some minutes (*cf.* Chapter IV., Section 2). Tyndall reached his method of sterilizing by the following train of thought: When a hay infusion has been boiled for some minutes all of the rod-shaped bacilli, unable to endure this treatment, are dead, while the spores of bacilli (more exactly, of some of them) are still alive.

If now all of these surviving spores are given time to germinate and the infusion is again cooked, it will be completely sterilized, because, when it is heated it contains no spores but only rods. However, as one cannot be sure that all spores have germinated at the same time, this process must usually be repeated several times. Hence it is recommended to submit the filled culture apparatus to steaming twice, with a day intervening. This is especially important for those culture media which, like gelatin, bear only a short cooking.

c. Waiting and Watching.—For greater certainty, the materials treated in this way are allowed to stand for some time before they are used, so that one may be sure that they contain nothing capable of germination. If they are placed in the thermostat at about 30° C., a delay of a couple of days may be looked on as sufficient.

3. *Water Bath for Sterilizing at Temperatures below the Boiling Point.*—Various fluids will not bear being sterilized at 100° C., because at so high a temperature they undergo chemical changes which it is wished to avoid for one reason or another. In such cases the short exposure to a high temperature can be replaced by long maintenance at a lower temperature, until no germ capable of growth remains in the fluids. This method was discovered by Pasteur, and by his advice applied on a large scale in the fabrication of wine ("Pasteurization") to secure sterilization of the wines without at the same time injuring their quality. In what follows we shall make use of the preparation of blood serum according to Koch's directions. Rather complicated and expensive apparatus is often used for this, but one can get along very well with a simple cylindrical water-bath made of tin, such as Koch himself used in his first preparations. The tin receptacle (Fig. 5) is 22 cm. high, and 13 cm. in diameter. At the top it is provided with a collar 1.5 cm. broad. A piece of fine-meshed flexible galvanized iron netting (*a*) serves as a lid, and is fastened by bending its edges under the collar. The temperature of the water is read off on the thermometer (*b*) which projects through a hole in the lid. The bulb of the thermometer is held up several centimetres above the bottom of the cylinder, by another loose piece of wire cloth (*c*) resting upon its edges, which are bent down; when the water has reached a suitable temperature, usually 60° to 70° C., it is easy to hold it at about

the same point for a long time, *e.g.*, a couple of hours, by the use of a small gas or alcohol flame. Though a range of a few degrees can usually be allowed, it is best to keep the temperature under observation. Further directions for the preparation of serum are given later.

4. *Porcelain Filters.*—Even a slight heating often causes in the culture fluid chemical changes which it may be important to avoid. In such cases the germs have been removed by filtration through burnt clay (Klebs), plaster (Pasteur and Miquel), porcelain (Chamberland), or sheet asbestos (Hesse),

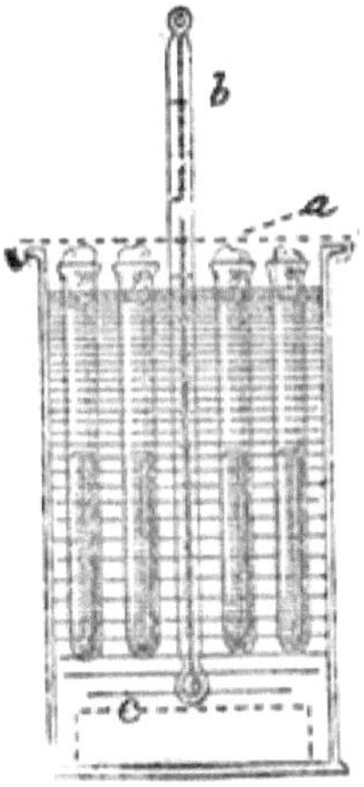

Fig. 5.

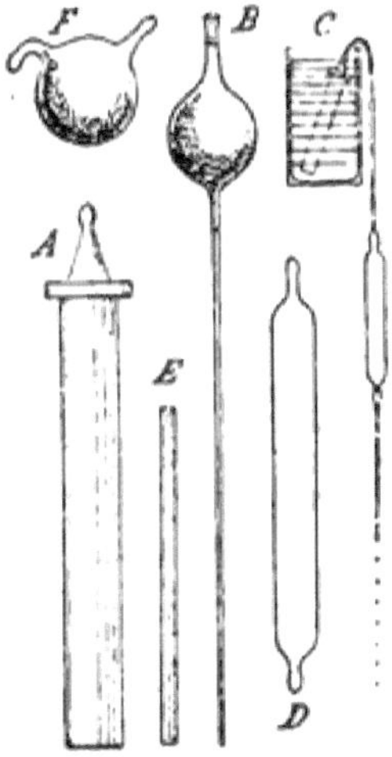

Fig. 6.

Fig. 5.—Water-bath for Sterilization at Lower Temperatures.

Fig. 6.—*A*, Chamberland Filter (× 1/6); *B*, bulb-pipette for collecting the contents of the filter; *E*, small porcelain filter (× 1/6); *F*, reservoir, which can also be used as a culture-vessel; *C* and *D* are explained in the text.

which keep back all of the cells contained in the fluid. As filtration plays a prominent part in the separation of bacteria from their soluble products, we shall here dwell somewhat more fully on the use of

Chamberland's Filter.—This consists of a hollow porcelain cylinder (Fig. 6 *A*), which is closed at the bottom and at the upper end provided with a funnel-shaped glazed end-piece, over which a rubber tube can be pushed. The filter is immersed in the fluid to be sterilized and the upper (open) end is connected with some kind of aspirator. If the fluid to be filtered is in too great quantity to be contained in the cavity of the filter, another receptacle is inserted between the aspirator and filter. An ordinary wash-bottle (Fig. 7) can be used for this purpose; and when the fluid to be sterilized is not too hard to filter, a common bottle aspirator (Fig. 7, *a*) can be employed. The details of the process are then as follows: The open end of the filter is closed by a plug of cotton batting. It is then wrapped in paper and sterilized at 150° C. The open ends of both tubes of the wash-bottle are plugged with cotton, wrapped in paper and sterilized with the filter, as is the plugged wash-bottle. The rubber stopper of the latter (bored

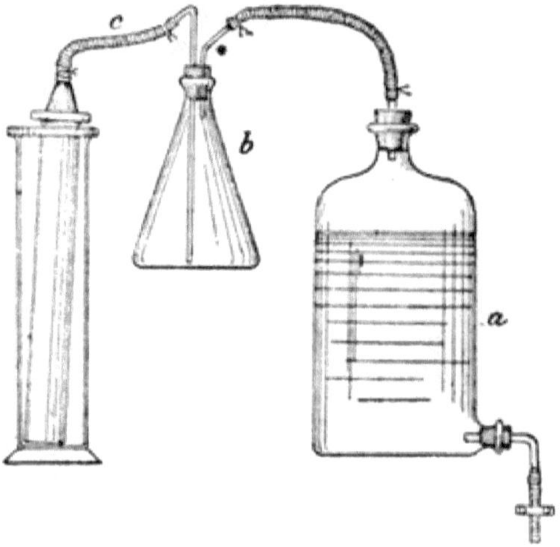

FIG. 7.—Chamberland Filter set up for use, with Aspirator and Wash-bottle Reservoir.

with two holes for the tubes), and the two pieces of firm-walled rubber tubing for connecting the flask with the filter and the aspirator, must be sterilized in some other manner. They are first laid for fifteen minutes in a $\frac{1}{10}$ per cent sublimate solution, rinsed in sterilized distilled water, plugged with cotton that has been previously sterilized by dry heat, and after wrapping in paper they are steamed for fifteen minutes. After this has been done, the filter is put together as rapidly as possible, and with the ulmost cleanliness, the fingers being first washed in sublimate. The glass tubes are fixed in the rubber stopper, and this is fitted to the wash-bottle after removing the plug from the latter with sterilized forceps. The long tube is then joined to the filter by one piece of rubber tubing after removal of its cotton plug. The only plug left in place is that at * of Figure 7. The entire apparatus is now steamed for ten minutes, after which it can be used. For additional safety, it is usually customary after the last sterilization to bind the rubber tubing fast to the filter and glass tubes by cord, or better still by previously glowed coarse brass wire, which is tightly wound by the use of pliers.

If instead of the common wash-bottle, a so-called Pasteur flask, made entirely of glass, is used (*cf.* Chapter XI.), the preparations for filtering are very much simplified. It is also evident that if an autoclave is at hand the preparatory sterilization, as well as the rapid and careful setting up of the apparatus can be dispensed with. Without any especial precautions the glass and rubber parts can be joined and then sterilized in a few moments at 120° C.

Figure 7 shows the apparatus in use. The filter is immersed in a relatively narrow glass cylinder, which contains the fluid to be filtered, as this sinks in the glass it is tipped so as to use all of the filtering surface and to prevent air from being sucked through, though this is also sterilized in the passage through the filter. The simple aspirator (*a*) is sufficient, when one has not to deal with fluids that are hard to filter. The bubbling of air back through the aspirator, and the consequent hindrance to filtration, is avoided by regulating the size of the escape tube by a pinch-cock. In other cases, or when the filtering must be done rapidly, more powerful suction is employed, *e.g.*, the small air pump recommended by Chamberland (Fig. 8, *a*), or a filter-pump worked by water. If the filtration is difficult and a powerful pump is used, care must be taken that the walls of the flask are strong enough to bear the atmospheric pressure. In such cases it is customary to use a thick-walled flask instead of a common wash-bottle, and under these circumstances exceptional use may be made of a lead tube within the rubber tubing. After repeated sterilizing by steam, even thick-walled rubber tubing becomes too much softened for further use.

When the wash-bottle (*b*) is filled with the clear filtrate, it is first loosened from the aspirator, then the tube (*c*) is fastened by a clamp or glass plug, after which the flask can be separated from the filter and its contents kept sterile. The filtrate remaining in the cavity of the filter can be saved, by pouring it into one or more sterile vessels, *e.g.*, culture tubes, as soon as the filter is opened. This is best done by using a pipette (Fig. 6, *B*), plugged at the top with cotton, while the tube below the bulb is long enough to reach the bottom of the filter and slender enough to pass through its opening. Before use, it is hermetically sealed, sterilized, and passed through the flame like an ordinary Pasteur pipette, of which, indeed, it is only an enlarged edition (*cf.* Fig. 22).

After being used, the Chamberland filter is cleaned according to circumstances, by simply brushing off the surface and washing it in water, or by the additional use of chemicals, *e.g.*, disinfectants. Occasionally it may also be desirable or necessary to rinse its pores out thoroughly with water, before it is dried and again sterilized. This is most quickly accomplished (Fig. 6, *C*), in the following way: the filter is filled with water and immersed in a large vessel of water, which is placed quite high, as on a shelf or cupboard. A rubber tube several feet long is attached to its open end. Somewhere in this a glass tube (Fig. 6, *D*), filled with water is inserted. Before putting together, the filter and tube are both filled with water.

When one has to do with small quantities of fluid the small thick-walled filter (Fig. 6, *E*) can be used. This is 15 cm. long, with a cavity 2 mm. in diameter. These were first used in Pasteur's laboratory, inserted in a filtering apparatus of special construction, but they can be used in the same manner as the larger ones, by fastening a rubber tube to the upper end with copper wire. A strong suction apparatus is needed for this. The small bulb (Fig. 6, *F*) is intended to be used as a receiver of filtered fluid (Fig. 8) and also without special change, as a culture vessel, with tubular plug (*cf.* Chapter II., p. 16).

It must be remembered that even filtration through porcelain does not always avoid the chemical modification of fluids, since the filter may hold back certain compounds, as well as the germs. If organic culture fluids must be absolutely unchanged, and at the same time free from germs, it is necessary to do without all sterilization, in which case they are to be collected or prepared with such care that all contaminations from their surroundings are avoided. Blood and urine, for instance, are drawn directly from the heart, veins, or bladder with antiseptic precautions. The preparation of originally sterile infusions is further briefly discussed in Chapter XI.

5. *Disinfecting Solutions.*—As a help in sterilization, it is usual to have at hand a large quantity of corrosive sublimate solution (1 or 2 : 1,000), which is not only used in preparing some sorts of culture media, *e.g.*, potatoes (see Chapter III.), but in many other cases, for cleansing the hands, instruments, morbid material, etc. If well water is used instead of distilled water in making this solution, care must be taken that no precipitate of insoluble mercurial salts is formed, or the fluid

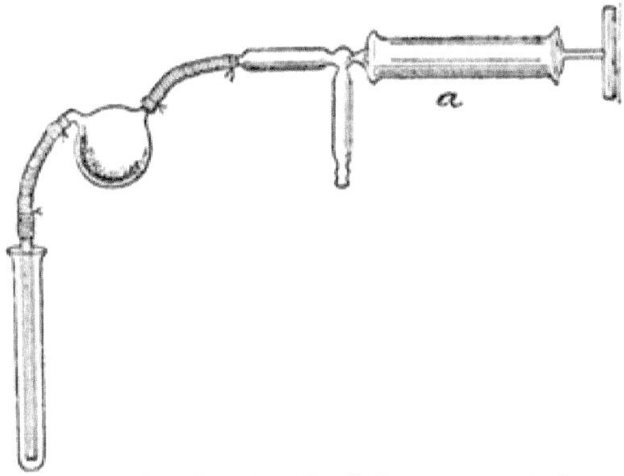

FIG. 8.—Small Porcelain Filter joined to Bulb-reservoir and Air-pump (*a*).

will possess no disinfecting value. To prevent this, a small quantity of acetic acid may be added to the well water. Fürbringer recommends 0.5 gram to each litre of 0.1 per cent solution. Carbolic acid solutions of various strengths are also to be kept ready for use.

CHAPTER II.

COMMON CULTURE-APPARATUS.

THE principal requirements of a culture-apparatus in which pure cultures are to be conveniently and surely obtained are the following: Besides being durable, simple, handy, and cheap, it must be easily sterilized with its contents, and so that it can be opened for inoculation and again closed without much danger of infection from the air. It must also be so closed that the entrance of foreign germs is impossible, while free access of the air is permitted (except when anaerobic forms are to be cultivated). This is effected by the use of cotton batting, which allows a sufficient circulation of air while filtering the germs from this.

The apparatus to be recommended for simply keeping a pure culture is the following:

1. Test-tubes and flasks plugged with cotton. One can commonly use:

a. Test-tubes of any size. The use of small tubes (*e.g.*, 13 cm. long, and 1 cm. in diameter) is to be recommended. A considerable quantity of the culture material is then saved. Since the heating of these glasses is generally effected in the steam sterilizer and not directly over the flame, their walls need not necessarily be so thin as those of chemical test-tubes, and consequently it may be best to order those with especially thick walls, entirely without a flange, so that the cotton plug may fit better to the edge.

The test-tubes are first carefully cleaned with common distilled water, and then plugged in the following manner: Good cotton batting (not absorbent cotton) is pressed with a pair of forceps (or twisted by the hand) into the mouth of the test-tube in sufficient quantity to form a firm, tightly fitting plug, which reaches a couple of centimetres into the tube, while part of it projects and frays out over the edge. The plug must be both firm and tightly fitting, yet capable of removal and reinsertion (by a twisting motion) without much difficulty. When the test-tubes have been plugged, they are put in a small four-sided basket, made of wire-cloth, of a suitable size for the oven, and sterilized at 150° C.

If it is wished to make use of a larger surface than is available when a test-tube is used, even when it is obliquely placed (Fig. 9, II.), recourse is had to:

b. Small Conical Flasks.—(Erlenmeyer flasks) of the form shown in Fig. 9, VII., and holding about 100 cc.

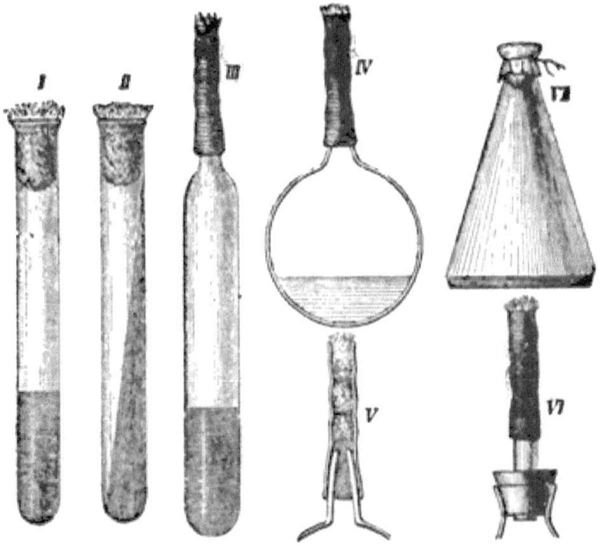

FIG. 9.—Various Culture-glasses with Cotton Plugs (*I., II., VII.*), or Plugged Caps (*III.—VI.*)

c. Small Medicine Bottles (30 gm.) are conveniently used in the same way, and the surface of the gelatin in them can also be increased by placing them in an oblique position while it is hardening. They have the advantage over test-tubes that they may be stood upright without the use of a rack.

Any of these receptacles can be used for cultures in either solid or fluid media. A fault common to them is that they are all relatively wide mouthed, so that the removal of the plug leads to a certain danger of contamination, especially from dust which has settled on the plugs when they have stood for some time. This can be obviated by tying a couple of layers of filter paper over the cotton plug (Fig. 9, VII.), which is especially recommended when large flasks are used. For greater safety, the part of the cotton projecting beyond the tube can be singed off just before use. Both these precautions are less necessary when solid media are used, so that the tube or flask can be inverted while the plug is removed and inoculation effected. They are more important when the contents of the vessels are fluid. Much greater safety is secured by the tubular cotton plug of the author which is described later.

2. *Vessels with Tubular Plugs.*—When I began in 1879–80 the pure culture of a large number of putrefactive bacteria, which had been isolated by the capillary-tube method, I first used stoppers in the form of short pieces of rubber tubing, in one end of which small sterilized tampons of cotton were fixed. The vessels which were closed by these tubular stoppers were larger and smaller flask-shaped or test-tube-shaped glasses, which had a short and broad slightly conical neck, with relatively narrow mouth (2 to 4 mm.) over which the rubber tube could be slipped (Fig. 9, III.–V.).

The tubular stoppers are prepared as follows: The cotton is sterilized at 150° C. The rubber tube, 6 to 8 cm. long, is sterilized by steam, wrapped in paper. It should be a little larger than the top and a little smaller than the bottom of the conical neck it is to fit. After cooling, the tubes are unwrapped, and by the aid of small forceps, half filled with tampons of the sterilized cotton, which must be so large and firm that they bulge the tube slightly (Fig. 9, V.). The advantages of these stoppers are, obviously: *a.* In opening and closing the flasks the cotton and the dust that has collected on it are not touched. *b.* The apparatus is opened and closed at a point where its surface can always be easily freed from dust. *c.* The opening through which inoculation is effected is smaller than with test-tubes.

The same principle was afterward applied in a more adequate form in the so-called Pasteur or Chamberland flasks (Fig. 10) which are much used in Pasteur's laboratory. The cap of these is not made of rubber but of glass, and is closely fitted to the neck by grinding. Unlike the flasks with rubber tubes, they can be sterilized entire at 150° C., and hence, as a rule, are preferable to the latter. The flasks and caps, which belong together, are numbered correspondingly with Brunswick black (Fig. 10).

Notwithstanding the decided superiority of the tubular cap over the simple cotton plug, the latter as a rule is used even in fluid cultures; but the rubber cap is always useful when a narrow tube is to be plugged with cotton (*cf.* Fig. 56 and Fig. 58). By means of a bored rubber stopper, a tubular cap can be adapted to any flask (Fig. 9, VI.).

3. *Other Culture Apparatus.*—A pair of small glass trays (3 to 5 cm. in diameter) can be used for cultures, one serving as a lid for the other (Fig. 13). They have the advantage over

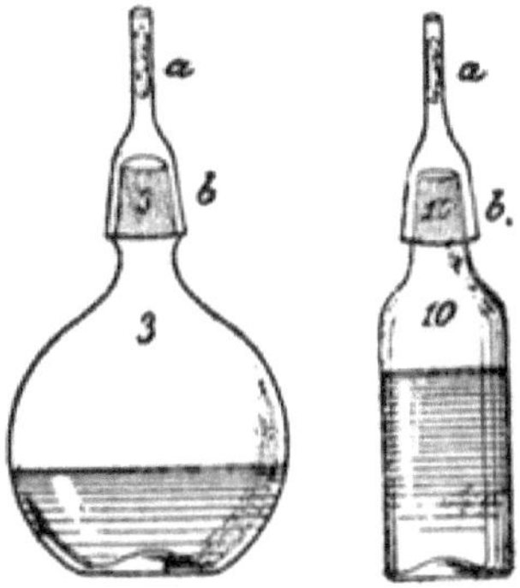

FIG. 10.—Pasteur-Chamberland Flask.

FIG. 11.—Glass Box for Soyka's Museum Cultures.

vessels plugged with cotton, that their contents are more easily accessible, even for microscopic examination, but they are less secure against foreign germs.

A far more reliable security against contamination is afforded by glass boxes with overlapping and ground covers (Fig. 11), such as Soyka uses for his hermetically sealed "museum cultures" (*cf.* Chapter VI.).

A number of other contrivances for cultures adapted to particular purposes will be described later. The apparatus for cultivating anaerobic bacteria, and for cultivating bacteria under the microscope, are for convenience left for treatment in later chapters, devoted to these subjects.

CHAPTER III.

CULTURE-MEDIA, AND THEIR INTRODUCTION INTO TEST-TUBES, FLASKS, ETC.

It would be useless to mention the enormous number of solutions, infusions, etc., which have been applied advantageously in various bacteriological investigations, and moreover it may obviously be necessary for each experimenter to use the most dissimilar culture-media, and to vary them, within wide limits, by the addition or removal of one substance or another according to the object of the investigation. Directions are given here for the preparation of only such fluid and solid media as have been found especially useful in the study of pathogenic micro-organisms during the last few years. They are commonly made in the conical Erlenmeyer flasks (Fig. 10, VII.). These conical flasks will be frequently mentioned in the following pages, where they are designated as large, medium, and small, holding respectively 1,500, 500, or 100 gm.

A. Fluid Media.

1. *Flesh Water*.—Even the simple decoction, made by cooking meat in water, with an acid reaction, affords a good nutrient substance for a large number of bacterian forms, but when it is neutralized by the addition of a little sodium carbonate, a fluid is obtained in which a large number of the most important pathogenic germs yet known, also thrive. For the preparation of broth, different kinds of meat are used according to circumstances, and the procedure is somewhat different in the various laboratories. This is worth noting, because we must be prepared to recognize in our present imperfect knowledge concerning the nutrition of bacteria, that slight variations in the preparation may give rise to differences in the results of cultures.

The two most common methods of preparation are the following:

a. Bouillon (B).—A pound of lean, scraped, or finely chopped beef, is placed in an enamelled kettle, or a large flask, with a litre of distilled water. The mixture is cooked for half an hour and filtered. The filtrate is neutralized or rendered slightly alkaline by adding to it, drop by drop, a solution of carbonate (or phosphate) of sodium. It is then again boiled for about an hour, by which time the insoluble albuminoids are all coagulated (C. Fraenkel), after which it is allowed to cool, when the fat solidifies. Having been once more filtered, the clear broth is filled into small vessels (usually medicine bottles [or test-tubes], plugged with cotton, *cf.* p. 21), in which it is finally sterilized in the steam-cylinder for a quarter of an hour. For greater security, it is again sterilized on the following day (disconnected sterilization, *cf.* p. 7). With each new cooking, the clear filtered broth may become clouded, but the turbidity disappears on cooling, otherwise it is necessary to repeat the cooking and filtration several times, until the fluid is obtained perfectly clear. This decoction of meat is known in the laboratory as bouillon (B) in contrast with the following.

b. Flesh water (K), in the more restricted sense. A pound of chopped lean beef is covered with a litre of water and set in an ice-chest for twenty-four hours, after which it is thoroughly shaken and filtered through muslin, the juices being well wrung out from the meat. In this way a litre of flesh extract is obtained, which is then cooked, filtered, rendered slightly alkaline, etc., as before.

The addition of 0.5 per cent of table salt increases the value of broth and flesh extract as a culture fluid for a number of bacteria (Miquel).

The addition of five per cent of glycerin (before the last neutralization) gives an excellent medium for the tubercle bacillus (Roux and Nocard).

Obviously, there may also be occasion for varying the composition of the bouillon by the addition of various other substances, *e.g.*, peptone, cane or grape-sugar, acetic acid, etc.

A useful flesh-water can also be obtained by the solution in water of a suitable quantity of some meat-extracts, followed by very careful "disconnected" sterilization (because of the numerous resistant germs often present in these extracts), neutralization, filtration, etc.

c. Liebig's extract (E), 5 gm. to a litre of water, is to be recommended, and especially

d. Cibil's extract (C), 20 gm. to a litre of water

Of other nutrient fluids, only the following, of proved and recognized value, will be named:

2. Aqueous decoction of liver, lungs, and other viscera.

3. Neutralized or slightly alkaline wine (used at one time by Pasteur) in the pure cultivation of the bacillus of splenic fever, on a large scale.

4. Infusion or decoction of wheat, hay, cabbage.

5. Yeast-water, a filtered and sterilized decoction of 100 parts water to 10 parts compressed yeast.

For the cultivation of yeast and moulds, the following are especially adapted:

6. Beer-wort, a decoction of dried and pulverized malt, obtainable at every brewery. This must be cooked for an hour and then filtered, before use, but it is hard to obtain it clear.

7. Decoction of horse dung (and of the excrement of other herbivora). One part of fresh horse dung is mixed, by the use of a glass rod, with three parts of water, set in a cool place for twenty-four hours, after which the mixture is cooked for an hour, and filtered through a double filter, which is a very slow process (unless a filter pump is used). The filtrate is again cooked for some time and if necessary refiltered, after which it is filled into small receptacles, sterilized in the steam-cylinder, and boiled twice for fifteen minutes, with a day between.

8. Decoction of prunes, which is best prepared as follows: The prunes are allowed to stand for a day in a little water, in which they are then cooked carefully, so that they remain unbroken. The fluid is afterward filtered and boiled down somewhat. In some cases it may be desirable to reduce the acidity of the decoction by the addition of sodium phosphate.

9. Decoction of other dried fruits, *e. g.*, raisins, dried pears, etc.

The horse dung, prunes, raisins, etc., used in making these decoctions, may be preserved in a sterile condition, for use as solid media in the culture of moulds, yeasts, etc.

A neutral or very slightly acid beer-wort is especially

adapted to the culture of Mucorini; while for the various species of Aspergillus a simple acid mixture of wort and prune-juice is an especially good medium, as O. Joh. Olson has told me. A mixture of the three liquids numbered 6, 7, and 8, may at times be useful. The reader is referred to Chapter XI. for directions for collecting and preparing primarily sterile nutrient fluids.

When a large quantity of culture fluid has been prepared, it is distributed in several medicine bottles (or Erlenmeyer flasks) holding one or two hundred grams, previously plugged with cotton and sterilized at 150° C. In these smaller vessels the fluid is finally sterilized in the steam cylinder for five to fifteen minutes on two sucessive days. It can then be preserved as long as is wished, in a dry place, provided the flasks are well plugged. The safest plan is to use a large cotton plug, and to tie over it several layers of filter paper (or to cover it with a sterilized thin rubber cap, such as the Germans now use extensively).

Formerly "Pasteur's fluid" (pure rock-candy, 10 gm; ammonium acetate, 0.1 gm.; and the ashes of 1 gm. yeast, all dissolved in distilled water), "Mayer's fluid," or "Cohn's fluid" (potassium phosphate, 0.5 gm.; magnesium sulphate, 0.5 gm.; tribasic phosphate of potassium, 0.5 gm.; acetate of ammonium, 1 gm.; water, 100 gm.) were used. These fluids are little suited to the cultivation of bacteria. It must also be remembered that Pasteur by no means introduced such mineral solutions as suitable nutrient fluids for microbes, but to show (1858) that yeast cells can produce albuminoids from a carbo-hydrate and an inorganic nitrogenous compound, when the necessary ash-constituents are also present.

B. Solid Media.

The systematic use of solid nutrient media, especially of nutrient gelatin (Robert Koch, 1881), marks a turning point in the history of bacteriological technology. Brefeld had previously employed nutrient gelatin, but essentially only to check the drying of slide-cultures. The starting point for Koch's important discovery was the long-known fact that the cut surface of a cooked potato, laid away for some time exposed to the air, becomes the seat of large and small colonies of mould, yeast, and bacteria, the last two of which often occur as small distinct, slimy colonies of various colors. Each colony, as a rule, contains only one form of yeast or bacteria, which has developed from a germ that fell from the air and found in the potato a favorable soil for its growth (first noticed by Hoffmann, in 1869, and utilized by Schroeter in 1872, in his cultivation of pigment bacteria). If we imagine these germs to have fallen, not on the solid surface of the potato, but on an equally large surface of some fluid in which they could thrive as well, it is easily seen that the several forms would have run together after a short time, motile and quiescent being mixed together. Some of the germs which succeeded in developing upon the potato, where they found space undisturbed by other colonies, would, perhaps, have failed to develop at all in the fluid, yielding to others in the struggle for existence. The same germs which in the fluid produced a motley tangle of intermingled forms, gave a series of well-separated colonies, on the solid medium.

When Koch had become aware of the extraordinary advantage offered by cultures on solid media over those in fluids, he sought to give various useful culture fluids a solid form, and he succeeded in doing this by gelatinizing them, in the manner to be described. The "nutrient gelatins" so prepared, have the advantage over potato, that their chemical composition can be varied within wide limits, so that solid culture media may be produced for such bacteria as cannot be found on potato. They are, further, transparent, which renders possible the observation of the growth of bacteria within the gelatin, as well as the microscopic examination of the culture. Finally, they are liquefiable at a low temperature, which is of decided value for their application to the isolation of the different bacterian germs.

A low melting point, however, limits in some ways the usefulness of gelatins, since it makes it impossible to employ them for cultures at much above 20° C. Koch therefore introduced for such cultures a second gelatinizing substance, agar-agar, which remains solid at the highest temperatures used for culture investigations. The same remark applies to sterilized blood-serum, which Koch found a method of preparing.

It was said above that the introduction of solid culture-media marks a turning point in the history of bacteriological technology. This is doubly true. In the first place, it has become possible by them to surely and easily isolate, and cultivate in a state of purity, the various bacterian forms, as is evident from what has been said, and as will be shown more in detail in the next chapter. But the introduction of solid media has also in many ways simplified bacteriological work, making it possible to work with far simpler apparatus, and at the same time with far better control and far greater certainty, aside from other reasons, because any accidentally introduced foreign germ manifests itself more readily when it forms a limited colony on or in the gelatin, than when its progeny in a fluid become scattered and mixed among all the other bacteria.

It must not be forgotten, however, that there are also limits to cultivation upon solid media, and that it by no means necessarily renders fluid cultures superfluous; so that the same care is due to the technique of the latter as formerly. To mention a single instance of many: it is obvious that experiments concerning the fermentation products of bacteria, and their nutrition, may demand the use of culture fluids which have a very simple chemical composition, but which would be changed into very troublesome and complex mixtures by the addition of gelatin or agar-agar. In addition to this, the usefulness of a culture fluid for certain bacteria is at times lessened or destroyed by the addition of gelatin, etc.

1. *Boiled Potato.*—According to circumstances, the preparation and application of potatoes varies a little, so that one either:

1. Simply cuts them in two, and lays them in a moist chamber. The problem in preparing them is first and chiefly, to remove or sterilize the dirt adhering to them, which always contains large numbers of particularly resistant bacillus spores. To this end as clean and smooth-skinned potatoes as possible are selected. Then they are repeatedly washed in water, and any remaining dirt is carefully scrubbed off under water, with a brush. By the use of a pointed knife any diseased or injured parts of the skin, as well as the changed parts of the underlying parenchyma, are removed. The potatoes are then laid for some time in a 0.1 per cent. sublimate solution, wrapped singly, without being dried, in thin wrapping paper, and ex-

posed for half an hour to streaming steam at 100° C. After twenty-four hours the steaming is repeated for fifteen minutes and the potatoes are ready for use. They are taken from the paper one at a time, held between thumb and one finger of the left hand, which has previously been washed in sublimate, and are halved by a table knife that has been carefully sterilized by being drawn several times through the flame, or by prolonged heating at 150° C. (wrapped in paper in the sterilizing oven). The halves are then quickly laid with the cut faces up, in a moist chamber, *e.g.*, under a bell glass set on an earthen plate (or in the flat pairs of trays similar to Fig. 13, but about 20 cm. in diameter which, though they cannot be so easily obtained, are safer and have the advantage that they can be set away one on top of the other). Dish and bell glass have first been carefully cleaned in water and rinsed in 1.0 per cent sublimate; and one or two thicknesses of filter

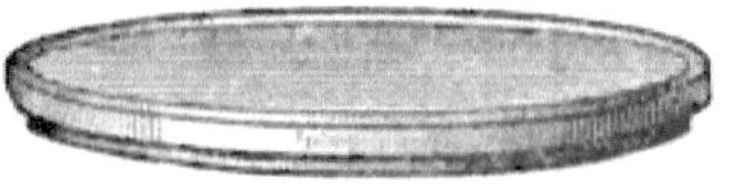

FIG. 12.

FIG. 13.

FIGS. 12 and 13.—Shallow and Deep Glass Trays in Pairs, the Larger serving as a Lid for the Smaller.

paper moistened in the same solution have been laid in the bottom of the plate. A potato cooked and divided in this manner, placed under a bell glass is one of the simplest arrangements for cultivating bacteria, but it is always exposed to contamination by germs from the air. This is avoided by:

2. Dividing the potatoes into prismatic pieces (or cylinders, by the use of a small-sized tin cutter made like common apple corers) and putting these in plugged test-tubes. In this case it is only necessary to cleanse the surface of the potato with brush and sublimate solution. After the potatoes have been steamed once for five minutes, they are pared and with sterile instruments cut into pieces which are quickly placed in the test-tubes already sterilized in the usual way, after which they are steamed for a quarter of an hour.

3. Nicer cultures are obtained by cutting the cooked and pared potato into round discs, which are laid in the bottom of a small pair of glass trays (Fig. 13, *cf.* Fig. 11, as well as Soyka's museum cultures), which after being wrapped in paper

are then steamed for fifteen minutes. Such discs are best cut out by aid of a small tin ring.

4. Potato broth prepared by mashing pared boiled potatoes and adding a proper amount of water can be occasionally used with advantage, if an especially large culture surface is desired.

2. *Gelatinized Media.*—If we desire, according to Koch's directions, to change our different culture fluids into solid and transpsarent but liquefiable substances we make use of the following:

a. Gelatin.—The finest French gelatin, which comes in thin oblong sheets of about 2.5 gm. weight, is used. Five, or commonly ten, per cent of this is added to the nutrient fluid, dissolved, cooked, rendered slightly alkaline, cleared, filtered, poured into smaller vessels and steamed twice with a day's interval.

b. Agar-agar (The Asiatic name of several peculiar gelatinous algæ, which grow in the Indian Archipelago and come into the market, dried in yellowish cartilaginous strips [or spongy prisms]).—When cooked in water this forms a stiff jelly and can be added to the various culture fluids in a quantity of 1 to 2 per cent, precisely like gelatin. As a rule we use 1.5 per cent, which is dissolved by long cooking and rendered slightly alkaline, after which it is cleared, filtered, etc.

In passing to a fuller description of the preparation of the gelatinizing substances, we must dwell briefly on the advantages and disadvantages attending the use of each. Agar-agar was introduced by Koch, as above indicated: (*a*) because it melts at a much higher temperature than gelatin, and can therefore be used for cultures on a solid medium at a higher temperature (30°-40° C. or higher); (*b*) it has also the advantage as compared with gelatin that it endures cooking for a longer time without diminution of its gelatinizing power; (*c*) there are many bacteria which liquefy gelatin in their growth, but do not affect agar in this manner, which in many cases is a great advantage, especially in isolation-cultures. On the other hand, it must be said: (*a*) that because of its low melting point, gelatin is better adapted to the isolation of germs; (*b*) it gives a filtrate as clear as water, while it is very difficult, not to say impossible, to get perfectly clear nutrient agar; (*c*) the difference between the mode of growth of different bacteria shows far more clearly in gelatin than in agar, so that two species which give in gelatin colonies of very different appearance, sometimes appear identical when they have grown in agar.

c. Agar-Gelatin.—To Jensen is due the credit of combining the good qualities of both media, and avoiding their disadvantages, by adding to culture fluids 5 per cent of gelatin, and 0.75 per cent of agar. The introduction of this mixture marks a real advance, and it is worthy of use as almost the chief culture-medium, since it is easily filtered clear, and liquefiable at so low a temperature that it can be used without difficulty for plate-cultures, though it remains solid at 30°–40° C.

d. Irish-Moss (Chondrus crispus), rarely used.—Neisser recommends a strength of 2.5 per cent.

The preparation of gelatins, agars, and agar-gelatins, is effected essentially in the same manner, so that a single description will suffice for all. But the length of time during which the solutions can be kept at the boiling point for sterilization, cleaning, etc., must be relatively short for those containing gelatin, which otherwise loses its power of solidifying. Agar, on the other hand, endures long cooking, while for agar-gelatin, a golden mean is kept. Gelatin, as a rule, is cooked ten minutes before filtration, and ten minutes after. Agar is best cooked forty-five minutes before, and a like time after; and agar-gelatin twenty minutes before and thirty minutes after filtering. Further than this, only clarifying and filtering demand special mention.

Clarifying.—When the nutrient jelly has cooked long enough, it is allowed to cool to somewhere about 50° C. An egg is then broken into 100 gm. water, and gradually added to the cooled but still fluid mixture, with which it is thoroughly incorporated. For a litre of jelly, the entire egg is used, a correspondingly smaller quantity being used for less than a litre. When the mixture is again heated to the boiling point, the white of egg is precipitated in a large yellowish curd, floating in a perfectly clear fluid. After cooking for some time, the next step is proceeded to—

Filtering.—This must necessarily be done while the jelly is still warm. If little is to be filtered, it is merely necessary to heat it up well before pouring it over the filter, so that, notwithstanding the cooling, it will remain fluid long enough for all to run through. But in this case, both the funnel and flask need to be first warmed. It is very easy to warm and sterilize flask, funnel, and filter, as well as to moisten the latter, in the manner shown in Fig. 14. A layer of water one or two centimetres deep is poured in the flask, and the funnel and filter are set in its mouth, the top of the funnel being covered with several thicknesses of filter paper, over which a plate of zinc or asbestos is laid (glass is apt to break). By heating the water to the boiling point for a few minutes, everything is sterilized, warmed, and moistened, in a single operation. While the water is still hot, it is poured out, and filtration can begin; the funnel being kept covered with a zinc or asbestos plate which prevents cooling to a considerable extent.

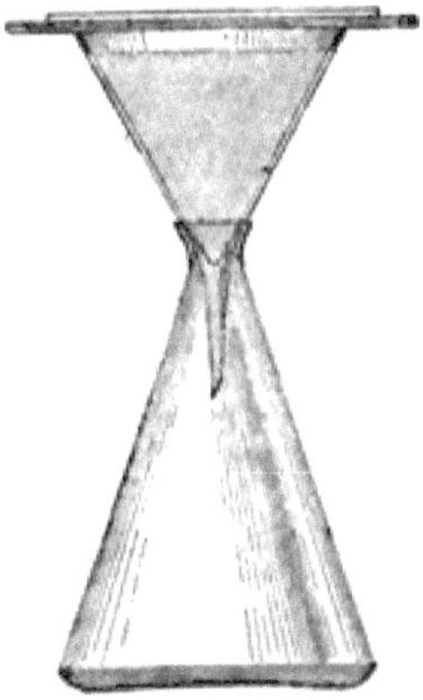

FIG. 14.—Simple Filtering Arrangement.

When larger quantities (*e.g.*, a litre) are to be filtered at once, there is danger that the gelatin may stiffen before the completion of the process, even though it was at first almost boiling. This is prevented by using a hot-water funnel (Fig. 15) a double-walled water bath with a projecting arm at *b*. The apparatus is filled with water through the hole *a*, and is kept warm throughout the filtration by means of a flame set under *b*. Such a double-walled funnel can be made by any tinner, and is far more convenient than the single-walled Plantamour funnel supplied by dealers in apparatus, since the latter must be plugged with a perforated rubber stopper at *c*, as the glass funnel limits the water on the inner side. To prevent too many germs from falling from the air into the flask, a little sterile cotton is stuffed into the mouth of the latter, about the tube of the funnel.

As examples, the preparation of the three most frequently-used nutrient jellies is given in detail. Other combinations (*e.g.*, C. A. G.) are also to be recommended.

C. G. *Cibil's Gelatin.*—20 gm. Cibil's extract is added to 1 litre of distilled water, in which 100 gm. gelatin is then dissolved. Heat till all is dissolved. Render slightly alkaline by addition of sodium carbonate. Boil for ten minutes.

Cool to 50° C. Clarify, as indicated above. Cook again for ten minutes. Filter through two thicknesses of paper in the hot-water funnel. Pour into smaller vessels, and sterilize for a short time in the steam cylinder on two or three successsive days.

E. P. A. *Peptonized Agar.*—5 gm. Liebig's extract; 30 gm. peptone; 5 gm. cane-sugar; 15 gm. agar; 1 litre distilled water. Cook for an hour, render slightly alkaline, and cool to below 60° C. Clarify, cook again for at least an hour, fill into bottles or test-tubes, and steam for ten minutes on each of two or

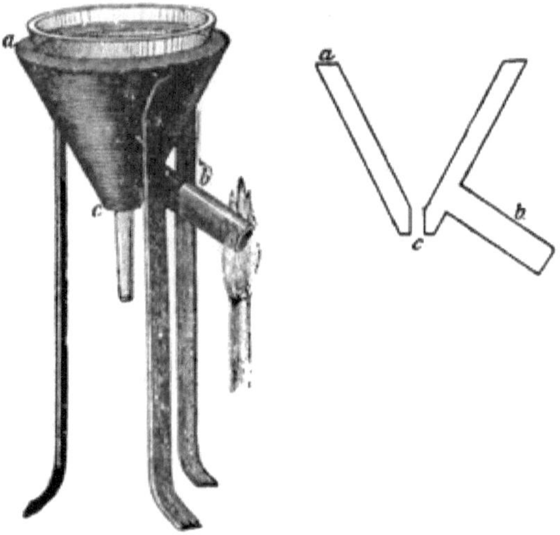

FIG. 15.—Plantamour Hot-water Funnel, for use with Gelatin and Agar.

three successive days. If 5 per cent of sterile glycerin is added to this agar, and the whole neutralized, it forms the glycerin B. P. A. recommended by Roux and Nocard for the cultivation of tubercle bacilli. This is far easier to prepare than serum, which is used for the same purpose, *cf. infra*, p. 29 et seq. and 32.

K. P. A. G. *Peptonized Agar-Gelatin.*—To a litre of filtered flesh-water (pp. 18, 19) add 5 gm. table salt, 10 gm. peptone, 50 gm. gelatin, and 7.5 gm. agar. Treat like E. P. A., except that it is to be cooked each time only 20 to 30 minutes. It is safest before final filtration to filter a little into a test-tube and see if it remains clear after boiling 10 to 15 minutes.

From my experiments, jellies made according to Mahn, by the use of two per cent of Cibil's extract (C. G.—C. A.—C. A. G.) afford quite as good a medium for many sorts of bacteria as the far more expensive peptonized flesh-water or meat gelatins, with or without sugar.

For the cultivation of moulds or yeasts, jellies are prepared by adding G. or A. to the nutrient fluids named above. Especially to be recommended, are:—

Beer-Wort Agar (B. A.), prepared of equal parts wort and water, with 1.5 per cent agar.

Raisin Gelatin (R. G.), a decoction of 250 gm. raisins in a litre of water, to which is added 10 per cent of gelatin.

3. *Serum.*—Some large glass jars (*e.g.*, pickle-jars—*cf.* Fig. 63) are tied up in three layers of paper, and sterilized at 140° C. They are filled with blood from oxen, calves, or horses (lambs' blood is not to be recommended), collected during the slaughtering of the animal, with as great cleanliness as can be obtained in a slaughter house. Care must be taken to avoid shaking the jars, and they are at once set away to coagulate, preferably in cold water. After this, they are kept in a cold place (best of all an ice-box), for 24 to 36 hours. The serum which has now separated out, is transferred in small quantities into plugged and sterilized test-tubes, by aid of a pipette that has been carefully sterilized. The greatest possible care is taken throughout, since any contamination will be more fatal here than in preparing the other culture media, because the final sterilization must be effected at a low temperature and consequently in an incomplete manner. Wetting the inside of the test-tubes toward the top with serum is especially to be avoided, as this part of the tube cannot be immersed in the water during the subsequent heating. The water bath (Fig. 4) is now filled with the test-tubes containing serum, and heated until the thermometer indicates 58° to 60° C., at which temperature it is kept for a little over an hour, by means of a small flame. This is repeated daily, for a week. The serum treated in this manner has become perceptibly clearer than it was originally, a small amount of whitish precipitate has separated, and a thin oily layer (of cholesterin) floats on the surface. All of the less resistant germs are destroyed by the heating.

It remains to solidify the serum without loss of its transparency. Koch effects this by prolonged heating at 68° C. The water bath is brought up to this point and carefully kept there during the process. According to the peculiarities of different lots of blood, the time required for solidifying serum varies from one to several (6 to 8) hours, so that it is necessary to examine a couple of tubes from time to time to see if the coagulation has begun. The heating is stopped when it becomes easy to invert the test-tube without loosening the serum, which is now an amber-yellow jelly, in all cases sufficiently clear in transmitted light to permit observation of the peculiarities of cultures growing along inoculation punctures made in it. The fewer red corpuscles originally present in the serum, the clearer and more attractive it becomes after coagulation; but use can be made of a somewhat red serum, especially for cultures on the surface.

The last heating at about 70° C. not only gelatinizes the serum, but also naturally contributes to a complete sterilization; yet one can only be sure of having killed all germs by convincing himself that the tubes remain free from bacteria after standing for some time, a couple of weeks, at the temperature of the room, or three or four days in the brood-oven at 30° C. (*Cf.* Chapter VI.)

An inconvenience in the use of the water bath for sterilizing and coagulating serum, already indicated, is that the cotton plugs and the upper part of the tubes, which project above the water, are not exposed to so high a temperature as the thermometer shows. This is avoided by using, instead of the water bath, a common hot chamber of the sort indicated in Chapter VI., such as it is always necessary to have for use as a brood-oven.

If it is desired to solidify the serum with a large oblique surface (Fig. 5, II.), the test-tubes must be laid in the chamber in a nearly horizontal position. They are best put in a flat box which can be obliquely set in the brood-oven.

Koch uses for this purpose a shallow, quadrangular thermostat covered with glass and felt. This is shown in section in Fig. 16. The bottom of this is 30 cm. square (inside measure) and the layer of water between the double walls is 6 cm. deep in the bottom (*aa*) and 3 cm. wide at the sides (*bb*). The projecting edge of the outside wall (*c*) rises a cm. above the rest and so incloses a space for a cover consisting of a square plate of glass (*d*) (which renders possible a quick observation of the temperature and the condition of the serum), and a sheet of felt (*e*) which is a poor conductor. Water is poured in through the tube (*f*), while air is allowed to escape through a similar tube at the other end. The apparatus is set obliquely by putting a couple of blocks (*g*) under two of its legs. This size accommodates two rows of test-tubes between which a thermometer is laid.

This combination of "Pasteurization" and "disconnected heating," which has been outlined (*cf.* pp. 7, 8) first used by

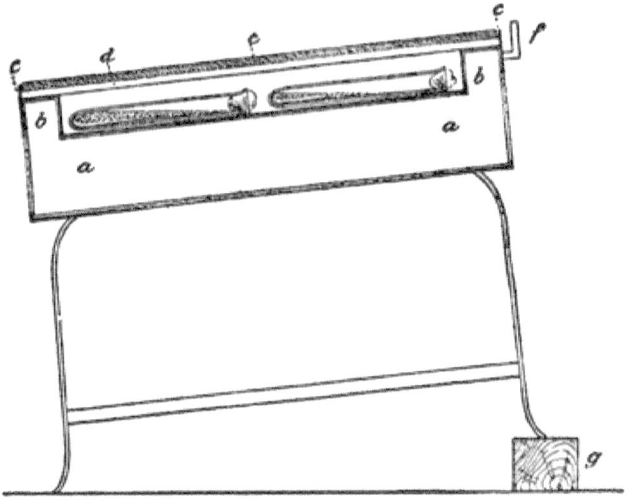

FIG. 16.—Shallow Thermostat for Solidifying Serum Obliquely.

Koch in preparing serum for the cultivation of the tubercle bacillus, must always be used when the blood has not been collected with the greatest possible cleanliness. But when the occasion offers for collecting blood by skilfully bleeding the animal, with sterile instruments, after carefully washing the neck of the horse or cow with 0.1 per cent sublimate, it is usually safe to omit the tedious and difficult preliminary sterilization and immediately proceed to solidify it at 68° to 70° C. after putting it into the test-tubes.

If it is desired to keep serum sterile for some time in a large vessel, the method that has been described cannot be used, but it is best to filter it through a Chamberland filter (under low pressure) into a Pasteur flask (Fig. 67), the tube of which is then quickly sealed by melting, and is first opened when the fluid is distributed into test-tubes.

Serous accumulations, such as those of ascites, pleuritis, or hydrocele, which can also be used as culture media (*cf.* Chapter XI.), are treated exactly like blood serum.

If 6 to 8 per cent of sterilized glycerin is added to the fluid serum and this is solidified at a slightly higher temperature than that which has been given (75° to 78° C.) solid glycerin serum is obtained, which, according to the observation of Roux and Nocard, is a much better culture medium for the tubercle bacillus than serum alone.

Serum can also be applied as a gelatinizing constituent of culture media, like gelatin or agar. Toeffler has advantageously used a mixture of three parts of sterilized fluid serum and one part of neutralized bouillon, containing one per cent of peptone, one per cent of grape sugar and 0.5 per cent of table salt. The mixture was solidified at 66° C.

4. *Softened White Bread.*—White bread is cut in slices and the soft part is broken from the crust in small pieces, which are laid in a thin layer under filter paper, to dry in the air. When, after a few days, it is dry enough to crumble, it is finely ground in a coffee-mill, and this dry bread powder is kept in a jar covered only with paper. When it is to be used, the desired quantity of the powder is weighed out, and poured into a sterilized flask, plugged with cotton, so that it forms a level layer in the bottom of this (about 8 gm. are used for one of the smallest Erlenmeyer flasks). By the use of a pipette, about two and a half times its weight of sterilized water is slowly poured over it. The bread broth formed in this way is sterilized twice by heating it half an hour in streaming steam, with an interval of a day.

White bread can also be used in slices which, either with or without previous sterilization by dry heat, are moistened with sterile water and then sterilized in the steam cylinder. But by pulverizing it, a more uniform and handy preparation is obtained.

This moistened bread is particularly adapted to the cultivation of moulds, but many different bacteria also thrive upon it. It can be improved and varied by moistening it with bouillon, prune juice, decoction of manure, etc. (Brefeld).

5. *Rice Milk.*—Very recently, Soyka has recommended the following mixture, which he has employed with success for his sealed museum cultures (*cf.* Chapter VI.): Rice meal, 10 gm.; milk, 15 cc.; neutral bouillon, 5 cc., mingled very thoroughly, filled into glass trays by a pipette, and sterilized by discontinuous heating in streaming steam on two successive days, during which it solidifies in the bottom of the tray as a white opaque mass.

6. *Other Culture Media.*—Of these, but three will be named: *a.* According to Soyka, other things being equal, spore formation in bacillus anthracis is hastened considerably by the addition of a certain quantity of sterilized sand to the nutrient bouillon; 2 to 4 cc. of bouillon to 25 gm. of sand gives a suitable degree of moisture to this "artificial soil."[1]

b. Cultivation upon moistened blocks of plaster was first used by Engel for inducing the so-called spore-formation in yeasts. Hansen recommends blocks in the form of truncated cones, 3 cm. high, 4 cm. in diameter at top, and 5 cm. at bottom, which are placed in small glass trays 5 cm. deep, loosely covered by similar inverted trays. Sterilized water is poured into the trays so as to reach to the middle of the block.[2] It is also possible to make these blocks of any desired form. I have, for instance, used successfully small plaster cylinders (moulded in glass tubes), that were placed in ordinary cotton-plugged test-tubes. It is to be observed that a mixture of eight parts of plaster and three parts of water is used in making the blocks. The mould in which they are cast must not be oiled. Before use they are sterilized by dry heat of 115° C., since a higher temperature robs the plaster of too much of its water of crystallization.

c. Eggs are shaken up so as to mix the white and yolk, and the surface disinfected with sublimate solution, after which a small hole is made through the shell at one end, by use of a needle, through which the inoculation needle is introduced, and the hole closed by a little cotton and collodion.

d. Colored nutrient gelatin. Noeggerath prepares the following mixture of concentrated aqueous solutions of aniline dyes, representing approximately the spectral colors: methylene blue, 2 cc.; gentian violet, 4 cc.; methyl green, 1 cc.; chrysoidin, 4 cc.; fuchsin, 5 cc., diluted with 200 cc. water. This is allowed to stand 10 to 14 days. Of the blue-black or dark gray fluid, 7 to 10 drops are added to each 10 cm. of peptonized gelatin (K. P. G.), which is cooked a couple of times, poured out on white porcelain plates, and infected by scratch-cultures (as in Fig. 29). The bacteria by their growth produce color changes in the gelatin, which can be used for diagnostic purposes.[3]

Filling the Culture Vessels.

So far as serum, bread, and rice-milk are concerned, the process has been already described. Only the gelatinizing media need closer consideration. These are stored, like culture fluids, in cotton-plugged medicine bottles holding 150 to 250 cc. each. It is convenient to have, in addition to these, smaller reservoirs, *e.g.*, large test-tubes, so that when only a little gelatin is needed it will not be necessary to open a large flask, since each time that one is opened, what remains in it must be re-sterilized for safety.

Fig. 17.—Pipette with Cotton Plug at *a*.

When the contents of a storage bottle are to be distributed into culture glasses they are melted in the water bath, or better, the steam cylinder. Test-tubes and other relatively wide-mouthed vessels can be filled by pouring directly from the bottle, or better, by the use of large pipettes (Fig. 17). These are cleansed, dried, wrapped in paper, and sterilized in the usual manner, if their size allows, in the dry oven at 150° C., otherwise they are drawn back and forth for some minutes through a gas or alcohol flame. To avoid contamination of the pipettes when they are laid down, they may be supported on small knife-benches (Fig. 18), such as are frequently used by housekeepers, the upper side of which is cleansed by flaming before they are used. The only other precautions to be taken are, to close the tubes as quickly as possible, and to see that each tube is filled for at least 3 cm. with gelatin, and that the top of the tube is not moistened where it comes in contact with the cotton plug, or the latter will become fastened to the glass. When flasks with a narrow neck are to be filled, this is best done by aid of a common wash-bottle, sterilized by steam.

After filling, the glasses are finally sterilized in the steam cylinder for five or ten minutes. If this apparatus cannot be used, each test-tube or flask is boiled by itself. When the gelatin has solidified, the test-tubes are best kept by wrapping a few together in paper, by which the surface, and especially the cotton, is kept from becoming dusty.

After standing for some time, the gelatin begins to dry out, as is shown by the sinking of its surface at the middle of the tube. When a needle is thrust into such gelatin, it does not close after the needle is withdrawn, but a crack forms, which may materially modify the appearance of the culture when bacteria eventually develop along the puncture. Before

FIG. 18.—Pipette on Knife-rest.

it is used, such gelatin should always be melted and allowed to solidify again.

To prevent the drying out of cultures on gelatin or agar, which is especially rapid when they are kept at an elevated temperature, the cotton-plugged ends of the tubes may be covered with small rubber caps. Care should be taken that both the caps and the cotton plugs are absolutely sterile, or there is danger of the cultures becoming contaminated. Mould spores, especially, which are unable to germinate in the uncapped and therefore dry plugs, germinate readily in the moist cotton, and grow through it. For this reason, it is necessary to singe the top of the plugs, and to carefully disinfect the rubber caps in sublimate immediately before putting them in place. The drying out of culture can also be prevented by singeing the cotton, and then dipping the top of the test-tubes in melted paraffin, which quickly hardens in an air-tight layer.

CHAPTER IV.

PURE MATERIAL FOR CULTURES.

IT has been shown in the preceding chapters that it is relatively easy to sterilize apparatus and nutrient media, as well as to keep them sterile for an unlimited time. The only serious difficulty in the cultivation of bacteria has always been connected with another point—the procuring of pure material for starting the cultures. The ubiquity and small size of bacteria have necessitated the application of quite peculiar and complicated methods. Their omnipresence makes it hard to avoid foreign germs, while their minuteness interferes with the sowing of a single germ, as is done in the cultivation of higher plants.

For certain forms of fungi this may be attained. Thus, Brefeld long since (1874) postulated and fulfilled the requirements for his mould cultures, that they should originate from a single spore, planted under microscopic control; and in 1883 Emil Chr. Hansen, in his work on yeast, which is so full of importance for breweries, prepared his cultures, which demonstrably start from one yeast-cell, by the use of other and better methods than those used by Brefeld.

On the other hand, the evanescent, small, and little characterized germs of bacteria cannot, as a rule, be handled in this way. It has been necessary to seize upon other means of getting pure inoculation material for cultures of these, and in 1881 Koch succeeded in a simple and ingenious manner in overcoming the difficulties and indicating a way of isolating bacteria germs, which will surely remain in the future as the chief method.

Before we proceed to a description of Koch's method, which, with various modifications, will be constantly used in the work that follows, we must briefly consider the earlier methods of securing pure cultures, not only because of their historical interest, but also because under certain circumstances, notwithstanding the discovery of the far more perfect process of Koch, we may be obliged to use them, or may find it advantageous to do so.

All of the methods which have thus far been used for securing pure inoculation material may be arranged in two principal groups: A, those in which use has been made of physiological differences between the bacteria for separating them; and B, those in which the actual separation of the germs from one another is used as the means of isolation.

A. Use of Physiological Differences.

If, for instance, one virulent form exists among a number of non-parasitic forms, in a putrid fluid, it may be obtained in a state of purity by the inoculation of a suitable animal, in the blood of which only the virulent form will come to development while the others die. In this way, from putrid blood, have been obtained the microbes of Pasteur's rabbit-septicæmia, Koch's mouse-septicæmia, and Davaine's rabbit-septicæmia. Or, for another example: according to Pasteur, in the blood of an animal dead of splenic fever, which has been kept at a high temperature, or which, from the size of the animal, has cooled slowly, there is found, so soon as putrefaction has set in, besides the anthrax bacillus, that which he has called the vibrion septique (which is identical with Koch's bacillus œdematis maligni). Pasteur succeeded in isolating these two pathogenic forms by utilizing their different behavior with respect to free oxygen; for as the first is anaerobic while the second is aerobic, and both grow in broth, he obtained a culture of one or the other by sowing the blood in receptacles from which all air was removed and kept out by carbonic acid (*cf.* Chapter VIII.), or in an ordinary culture vessel. These examples might be increased by many more, such as cultures in acid and alkaline media, with or without the addition of certain antiseptics, at higher or lower temperatures, etc.; but only two of the methods of this class will be dwelt upon a little more fully, viz.:

1. *Klebs' Fractional Cultures* (1873).—A fluid in which a large number of different bacteria live and thrive together, will necessarily not contain an equal mixture of all these

forms at every point. Some of the quiescent forms will sink to the bottom or adhere to the sides, while others can move about everywhere in the liquid. Some forms will form superficial pellicles, others thrive only below the surface, where there is less oxygen. Moreover, according to the nature of the fluid, some forms will fill it with great rapidity, and far surpass in numbers others which only manage to lead a restricted existence in the liquid. If a very small quantity of such a fluid containing bacteria is sown in a culture glass, in all probability only a few forms will be transferred, and these, less equally distributed, will develop in unequal numbers. If an extremely small quantity of this new mixture, containing fewer forms, is again transferred to a new culture flask, and this process is repeated several times, in all probability a point will finally be reached where the material used for the transfer is absolutely pure, *i.e.*, includes only a single kind of bacteria.

This method has only slight value, and it is especially to be observed that such fractional cultures are by no means to be counted on as finally giving a pure culture of any specified form among those that were found in the original fluid, not even that form which was most abundant there. The final product is often only one of the most widely distributed of the common putrefactive bacteria, which, even though originally present in small quantity, easily gains the advantage of the others. Still this method of fractional cultures may serve a useful purpose now and then, as supplementary to others.

2. *Cohn's Heating Method* (1876).—It had long been known that various organic fluids (milk, hay-infusion, infusion of peas, etc.), notwithstanding rather long cooking and the exclusion of germs from the air, became the seat of bacterian development, and adherents of the theory of spontaneous generation had repeatedly taken this fact as favoring their doctrine; but it was first shown by Cohn in 1876, in a series of experiments with hay infusion, that this must depend upon the extraordinary resistance of certain bacillus-spores to heat. The boiling heat rapidly destroyed all bacteria without spores, but the latter (in Cohn's experiments those of bacillus subtilis) where not killed—a fact which Brefeld also showed later by direct microscopic observation of the germination of the cooked spores. This afforded a means of obtaining pure material of certain spore-forming bacilli, and, as a rule, a pure

culture of bacillus subtilis can be obtained without difficulty in the following manner: A small quantity of water is poured over some hay in a flask, which is kept for hours at 30° to 40° C. (either in the brood-oven, or directly over the flame, while a thermometer is placed in the fluid). The dark reddish-brown infusion is diluted with distilled water until it reaches a clear golden color (s. g. 1.006), when it is filtered through muslin, neutralized with sodium carbonate, distributed in several vessels plugged with cotton, and cooked for ten minutes, after which it is set away in the brood-oven at 30° to 40° C. In the course of a couple of days, a larger or smaller number of the vessels will contain a growth of bacillus subtilis, which may be known by the fact that it forms a firm rigid film on the surface of the liquid.

It must be remembered that among the various resistant spores of bacilli there is a great difference in their ability to endure heat of 100° C. Some survive only a very few moments of cooking, while others bear this temperature for hours (*cf.* Chapter XII.). The limitations of this method are indicated by what has been said. It is possible to isolate by it only those bacilli which form very resistant spores, and a really pure culture is obtained with certainty only when the boiled fluid does not contain several different forms with the same power of resistance. Still there will be exceptional cases in which there is use for cooking as an aid in isolation.

B. Methods by Separation.

1. *The Capillary-tube Method* (Salomonsen, 1876).—Greater success has been reached by the use of methods in which an actual separation of the original germs is secured. The oldest of these is the following, the starting point for which was an observation of the color changes in putrefying blood. During decomposition, bright red ox blood, defibrinated by whipping, assumes a dark red or red-brown color, which is partly due to the removal of their color from the red corpuscles, partly to deoxidation of the oxyhæmoglobin, and partly to other chemical processes. If the blood, immediately after being defibrinated, is set away in a cylindrical glass where it can remain entirely undisturbed at a comparatively low temperature, *e.g.*, 10° C., it is seen that the change of color begins in spots here and there in the mass of blood. A closer investigation shows

that these putrefactive blotches are due to bacteria (or moulds or yeasts). Next the bottom (Fig. 19), where the crowded blood discs form a solid mass, in which the bacteria grow equally in all directions, the spots are round, clearly defined, and dark red. Above (*b*) they are elongated, often club-shaped, less clearly outlined, and by no means of so dark a color. Here the blood discs are suspended in a relatively large quantity of serum, and in this fluid the bacteria have been able to sink uninterruptedly to the bottom, marking their path by elongated spots or streaks.

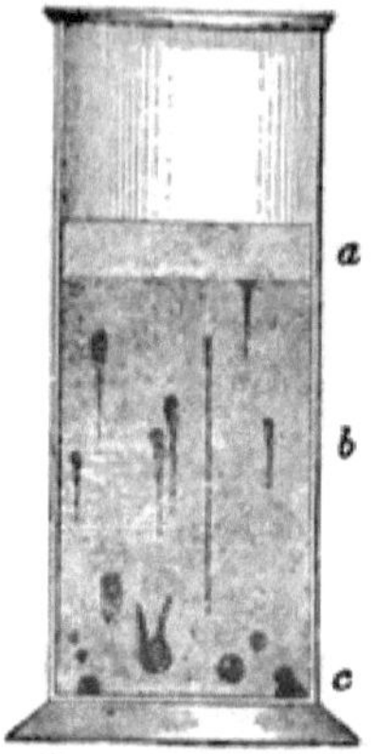

FIG. 19. — Cylindrical Glass Containing Whipped Ox Blood showing Purification Blotches. *a*, Serum layer; *b*, elongated discolorations; *c*, round spots and precipitated blood-corpuscles.

If this is to be utilized in obtaining pure material of the various putrefactive bacteria which grow in blood, it is only necessary to draw the defibrinated ox blood into long capillary glass tubes (50 to 60 cm. long and 0.5 to 1 mm. in diameter), and to attach these to strips of card-board about 3.5 cm. wide, by means of a drop of varnish at each end, which at the same time seals them. The following can be recommended as a suitable cement: 8 parts of rosin, melted with 2.5 parts of wax while the mass is constantly agitated. Of 1 part of turpentine, enough is added, little by little, so that a drop of the melted mass quickly solidifies when allowed to fall on a glass plate. Another good cement is the "Cire Galoz" (to be had of Alvergniat Frères, 10 Rue de la Sorbonne, Paris).

They are then laid away at the ordinary temperature of the room, for daily observation. The putrefaction blotches will soon appear; some shortly after the drawing of the blood, others many days later; some spreading with extraordinary rapidity, others growing only slowly. The number of each spot in sequence, the time of its appearance, as well as its growth from day to day, are easily noted upon the card. The growth is best indicated by pencil marks drawn every morning and night after the fashion shown in Fig. 20. This illustration, which shows one end of a card to which is cemented a capillary tube containing three putrefactive spots, renders further description unnecessary.

These putrefaction spots come from the bacterian germs which have chanced to get in the blood after its removal from the animal, so that in one lot there may be many of them, in another few; but each spot contains only a single sort of bacteria, developed from one germ. Consequently, when the discolorations are relatively far apart in the tubes so that they do not readily become confluent, each of them gives a small pure culture, from which pure material may be obtained of a specific form of bacteria. By this method, a mixed putrefaction-flora was first separated into its elements, so that the number of germs, the time of their development, rapidity of growth, etc., could be graphically represented.

2. *By Dilution* (Lister, 1878, Naegeli, 1879).—This method consists in diluting a fluid containing bacteria with sterile

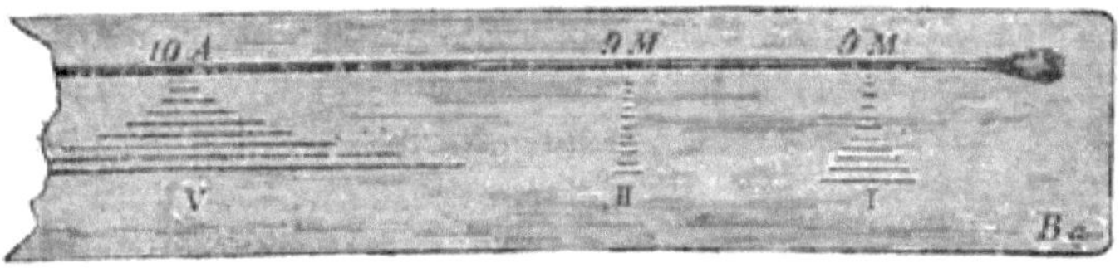

FIG. 20.—One end of a Pasteboard Bearing a Capillary Tube Filled with Blood. On the card are noted the time (10 A.—afternoon of 10th day; 9 M.—morning of the 9th day), when the spots appeared, the rapidity of their growth (indicated by the length of the lines), and their number in series (*I.*, *II.*, *V.*)

water to such a degree that a given quantity of the mixture holds only one germ, and in using such an initial quantity for inoculation. In the culture-flask, the offspring of this single germ forms a pure culture.

Counting the germs in the original fluid is effected under the microscope, by the aid of instruments similar to those used for counting blood-discs. From this the number of germs in, for example, 0.05 cc. of the fluid, is estimated, and by aid of a sterilized graduated pipette, this quantity is added to enough sterile water so that the mixture, after careful shaking, must average half a germ for each 0.05 cc. When by means of a sterilized graduated pipette, this mixture is inoculated into a series of culture-vessels, so that 0.05 cc., is placed in each, half of the glasses will not show any development of bacteria, having received no germs, while in all probability but one germ will have been placed in each of the others, which will then contain a pure culture. Lister devised and used this method for obtaining pure cultures of lactic acid bacilli. Naegeli, independently, cultivated micrococcus ureæ by aid of the method of dilution.

3. *By the Use of Solid (Especially Gelatinized) Media* (Koch, 1881).—This method consists in the isolation of the germs upon or in solid culture-media, where, in their later development, they give rise to separated colonies. If, in a flask like that shown in Fig. 9, VII., the bottom is covered with a thin layer of nutrient gelatin, and, after melting this at about 30° C., a very small drop of fluid containing bacteria is added and thoroughly distributed through the gelatin by careful shaking (*cf.* Chapter VII.), and after solidifying, the mixture is set aside at room temperature, after a time small colonies will appear at the surface of the gelatin. Each comes from one germ, so that pure material for inoculation is obtained.

This is but one of the many forms in which the method of Koch may be used. In a subsequent chapter, dealing with bacteriological analysis, the method will be treated particularly and in detail.

By aid of these methods, pure inoculation material can comparatively easily be obtained of all aerobic bacteria (*cf.* Chapter VIII.) which it is chiefly necessary to cultivate. It remains only to consider the best way of transferring this to the culture apparatus.

CHAPTER V.

INOCULATING CULTURES.

THE transfer of pure material to one of the culture vessels that have been described, is comparatively easy. The necessary opening of the different vessels, and the transportation of the inoculation material through the air necessarily entails danger of contamination. But this danger is very small, and if the work is done rapidly and carefully, the transfer is unsuccessful in a very insignificant number of cases.

The instruments needed are: 1, Platinum needles (Koch), long capillary tubes of glass, Pasteur pipettes, and glass needles.

1. *The platinum needle* consists of a glass rod about 27 cm. long and 4 or 5 mm. thick, in one end of which a piece of platinum wire about 3 cm. long is melted as shown in Fig. 21. Just before use, the surface of the glass rod is flamed, and the platinum wire brought to redness throughout its entire length. When, after a few seconds, it has cooled sufficiently, this is brought in contact with the material to be transferred, and quickly brought into the culture medium.

It is usual to have at hand some quite thin needles, and some which are a little thicker. An advantage of the former is that they cool quickly, while the latter are less flexible and can for this reason be used when the needle is to be thrust into more resistant substances, *e.g.*, liver or lung (or when it is necessary to apply considerable lateral pressure, as in distributing a colony of the tubercle bacillus over agar, etc.).

For certain cases, the platinum wire may be specially shaped. Small loops (Fig. 21, *b*) insure the adhesion of larger quantities of the inoculation material. A large loop (*c*), catches a good-sized drop. It is bent at a right angle (*d*), when one wishes to scratch the material into the surface of the gelatin (*cf.* what is said elsewhere about cultures on potato, slides, etc.).

When the inoculation must occur through a relatively small opening, through which the glass rod can be passed with difficulty or not at all, a sufficiently long piece of platinum wire can be used, or small pieces (5 or 6 mm. long) are employed by seizing them with sterilized forceps, glowing them, bringing them in contact with the inoculation material, and dropping them through the small opening into the culture flask. Capillary tubes and glass needles can, of course, also be used in such cases.

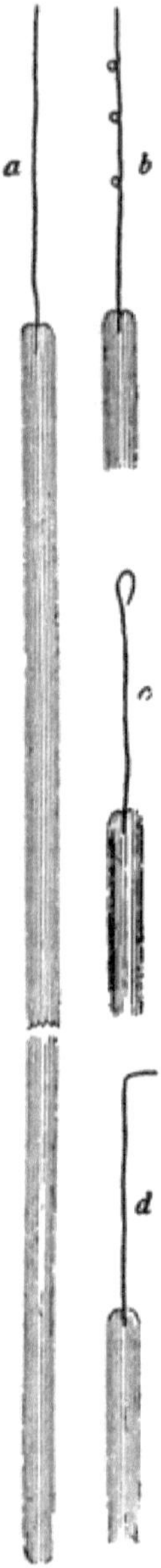

FIG. 21.—Platinum Needles, Straight and Bent in Various Ways.

FIG. 22.—A Pasteur Pipette.

2. *Capillary tubes* usually have a length of 20–25 cm. and are closed by melting at both ends. Before being used, one end is broken off, the surface is flamed, and the open end dipped into the fluid to be sown, when the other end is broken, upon which the fluid rises into the tube. The contents are blown out into the culture-vessel to be used; care being taken not to blow all out, but to leave a little fluid in the tube, thus preventing contamination from the air blown in.

3. *Pasteur pipettes* consist, as is shown in Fig. 22, of a piece of glass tubing, one end of which is closed by a cotton plug, which must not project beyond the glass, while the other end is drawn out into a capillary tube fused at the end. The plugged tubes are sterilized at 150° C. Before being used, the sealed end of the tube is broken off, and this part of the glass is flamed. The tube is filled by suction, and emptied by blowing the fluid out. These pipettes have the advantage over capillary tubes of being more capacious, and they

can be used as culture-glasses by reclosing the capillary extremity by heat immediately after filling. See also Chapter XI.

4. *Glass needles, i.e.,* glass rods drawn out into a very slender thread at one end for about 15 cm., are better than platinum needles in their perfect smoothness and rigidity, which may occasionally be of importance, as in making thrust-cultures of anaerobic forms, where it is wished to avoid introducing small air-bubbles into the gelatin.

One can easily make these inoculation instruments for himself by using an ordinary Bunsen burner or a good spirit lamp. A glass-blower's outfit naturally lessens the work, but it is not necessary.

For platinum needles, a glass rod about 4 mm. in diameter is divided into pieces 25 or 30 cm. long, by slightly marking it at the proper points with a triangular file, when by steadily pulling the glass in opposite directions at each side of the scratch, with a slight side motion, it is easily and evenly broken at the desired spot. The end of a piece into which the needle is to be fastened is rotated in the flame with the left hand, until it becomes red and soft, while a piece of wire of the right length, grasped some 5 mm. from one end in a pair of forceps, is brought to a white heat at this end and carefully thrust lengthwise into the softened glass, which is then allowed to cool gradually, being held for a few seconds close to the flame. The other end of the glass is finally rounded off by heating it to redness, rotating it meantime.

A glass tube 6 or 8 mm. in diameter is easily drawn into capillary tubes in the following manner: The tube is heated to redness for a short distance (1–2 cm.), while being rapidly revolved about its axis, until it becomes thoroughly softened, when it is removed from the flame and drawn into a tube about six feet long, which is melted off from the thicker glass at each end. In the same way, by aid of the flame, the long capillary tube is divided into pieces of the length indicated above, which are at the same time hermetically sealed at the ends. The strong heating of the glass, and the sealing of the tube which immediately follows, insure freedom from germs. The larger the original tube the greater the length for which it is melted, and the slower it is drawn out or the shorter the length of the capillary tube the coarser this will be, and conversely. This being borne in mind, it is easy to experiment until a tube of the diameter wished in a given case is obtained.

Pasteur pipettes are made by dividing a glass tube into pieces 15 cm. long, by aid of a file. Each piece is then heated in the middle while being revolved, and drawn out in the manner already described into a capillary portion about 30 cm. long, which is then melted and sealed at the middle, making two pipettes. The sharp edges of the other ends are rounded off by briefly glowing in the flame. After they are cooled, they are plugged with cotton, and finally sterilized at 150° C.

Glass needles are drawn out in the same way, solid rods being used instead of tubing.

To indicate the many little manipulations necessary in making a pure transfer, a detailed account is here given of (*a*) the inoculation of gelatin in a test-tube by use of the needle, and (*b*) the inoculation of fluid in a narrow-necked flask, by the use of small pieces of wire.

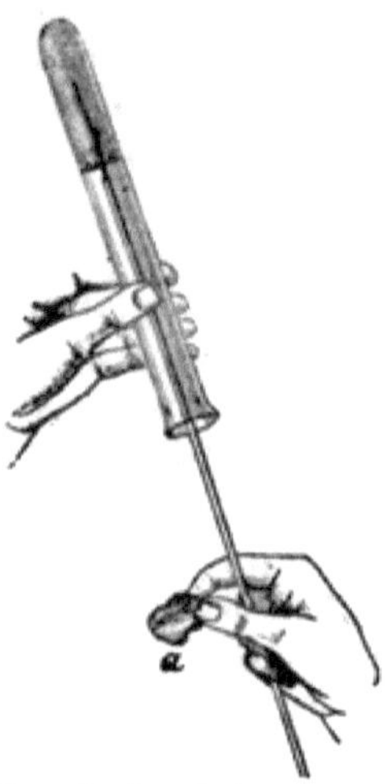

FIG. 23.—Inoculating a Test-tube of Gelatin.

a. Several tubes of gelatin that has not dried out too much (*cf.* p. 474) are chosen, and the cotton plugs are tested by twisting them several times to be sure that they are not glued fast to the glass. The needle is sterilized as indicated above. After allowing it to cool for a couple of minutes, the tube from which the transfer is to be made is opened by removing the plug by a twisting motion which causes any fibres that may have adhered to the glass to lie out of the way of the needle. The tube is held with the top upward, or inverted, according as the bacteria growing in it have liquefied the gelatin or not. The removed plug is held so that the part of it which comes in contact with the inner face of the glass shall not touch anything by which it might be contaminated. The needle is now quickly plunged into the culture, removed, and the plug replaced. (As a rule only the tip of the needle is brought in contact with the culture, a sufficient number of germs will always adhere to it. If for any reason a large quantity is to be transferred, loops such as are shown in Fig. 22, *b c* are used.) The tube to be inoculated, held in an inverted position (Fig. 23), is now quickly opened with the same twisting motion, the needle is rapidly thrust into the gelatin at one or more points, the plug replaced, and the needle at once sterilized by glowing. A label containing record of date, source of material used, etc., is then pasted upon the tube.

b. The surface of the flask is carefully dried, after which the lower part of the rubber tube and the contiguous part of the neck are flamed by use of a burner or alcohol lamp, and the rubber is slipped up a little, so that it can easily be removed by one hand. Sterilized forceps are placed upon glass benches (Fig. 18) or in a shallow glass. Small pieces of platinum wire are kept ready in a glass tray. One of these is grasped by one end with the forceps, at right angles to its axis. The wire is brought to a red heat, allowed to cool, and plunged into the colony of bacteria, etc., from which the transfer is to be made. With the left hand, the flask is now quickly opened, the wire dropped into the narrow opening, the cap firmly replaced, and a label prepared.

CHAPTER VI.

BROOD-OVENS AND THERMO-REGULATORS.

A MAJORITY of the bacteria known and cultivated up to the present time thrive at the ordinary temperature of a room (15° to 20° C.). Consequently if it is only desired to keep cultures going and to observe their growth, it usually suffices to set them aside in a living room. For preserving cultures for a long time, especially as museum specimens, Soyka recommends inoculating solid media in the small glass boxes shown in Figure 11, which are then carefully sealed with paraffin. By this means he succeeds in checking the growth of the colony at a certain point of its development, while the capability of germination is preserved, and drying out is hindered. For details see "Zeitschr. f. Hygiene," 1888. After a shorter or longer incubation period, they will be seen to start into growth, and even by the naked eye it is possible to observe the great dissimilarity which often exists between cultures of different bacteria. Even in 1880, when fluid cultures were almost exclusively used, I noticed in detail the microscopic differences that cultures often present, and called attention to their indications in judging of the purity of a culture, and the diagnosis of bacteria, which was sometimes more easily made by the naked eye than microscopically. When, later, Koch's gelatin cultures came into use, still more occasion was offered for the observation of evident microscopic differences between cultures of bacteria, such, for instance, as are shown in Figures 42 and 69. Sometimes it is very necessary to cultivate bacteria at a higher and more uniform temperature than that usual in our living rooms, for some forms develop quickly only at higher temperatures, others never thrive below 30° C., and some species demand heat for the production of spores. Moreover, indeed, as Pasteur first showed, cultivation at a high temperature may change the physiological character of bacteria, and is used in the fabrication of "vaccines." It may also be frequently desirable to cultivate a micro-organism for a long time at a constant temperature, whatever this may be.

Hence are used thermostats or brood-ovens, in which the temperature is held unchanged for weeks or months at the same

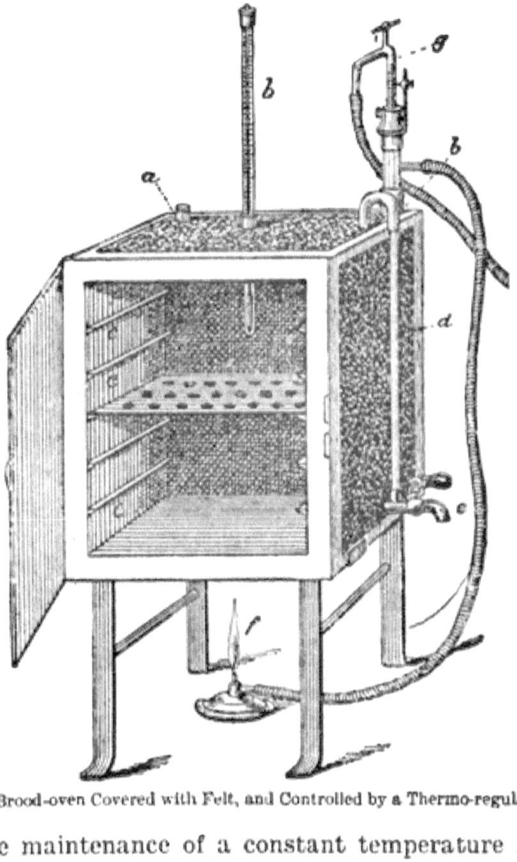

FIG. 24.—Brood-oven Covered with Felt, and Controlled by a Thermo-regulator.

point. The maintenance of a constant temperature is beset with certain difficulties. The changes in temperature of the room in which the thermostat stands is disturbing; to counteract them the brood-oven is inclosed in as thick a layer of some non-conducting material as possible, and set up in a cellar or similar place, where the temperature changes but little. Variations in the strength of the source of heat used also act in the same way, as is especially evident when gas is used, since the gas pressure may vary much in the course of the day. This may be obviated by the use of pressure-regulators and thermo-regulators of various construction.

A serviceable small thermostat, comparatively cheap, and self-regulating, is shown in Figure 24. It is a quadrangular box of zinc or copper, set upon a strap-iron support with legs about 24 cm. high. The top and bottom, as well as three of the side walls are double, and the space between them, which is 2 to 2.5 cm. wide, can be entirely filled with water through the opening (*a*), which is afterward closed by a cork. The fourth side is single, and serves as a door.. In the centre of the top, an open tube 3 cm. in diameter is fastened into the wall, piercing it. Into this by aid of a perforated cork, a thermometer (*b*) is set, which projecting into the interior of the box, can be examined without the necessity of opening the thermostat. Usually a thermometer is also kept entirely in the thermostat, standing in a vessel filled with oil, which prevents the mercury from sinking too rapidly when the glass is removed to observe the temperature. On the inner face of the two side walls are fastened a number of small ledges (*cc*) on which glass or tin shelves may be laid to provide support for a large number of low objects. The outside of the box is covered with felt, and provided with a glass tube for showing the height of the water (*d*), and a faucet (*e*). It is heated by a Bunsen burner, the tube of which has been removed, so that it gives a pointed white flame, which permits the flame to be reduced to a minimum with no danger that it will "snap back." [If the brood-oven is to be kept but little above the room temperature, there is some danger that the flame may occasionally go out, allowing the escape of a large quantity of gas into the room before it is discovered. For this reason, the safety burner devised by Koch, and to be had of dealers in bacteriological apparatus, though somewhat expensive, is to be strongly recommended, since it promptly and automatically cuts off the supply of gas in case such an accident occurs.—W. T.] The temperature is held at a constant point by the use of the regulator (*g*), which is passed into the water-filled space through an opening at *b*. Two generally used and good types of thermo-regulator are shown in Figures 25 and 26. Rohrbeck's (Fig. 25) is especially to be recommended.

Reichert's latest regulator is shown in Figure 26. The gas

enters at *a*, passes through the opening *b*, and out toward the burner at *c*. The column of mercury is adjustable by the screw *d*. In case this should rise so as to entirely close the opening *b*, the flame is not extinguished, for a minute opening is provided in the T-shaped tube at *e*, which permits the passage of just enough gas to keep the flame alive, without heating the thermostat appreciably. The right size of this pin-

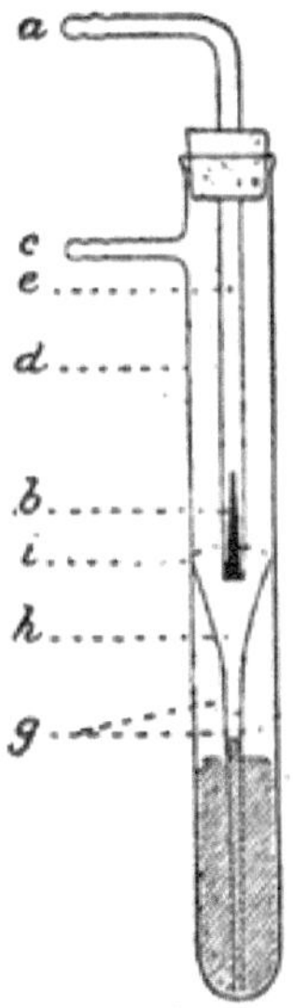

FIG. 25.—Rohrbeck's Thermo-regulater.

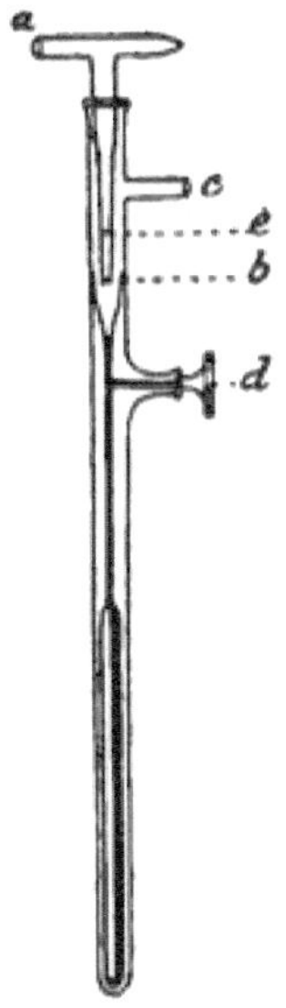

FIG. 26.—Reichert's Thermo-regulator.

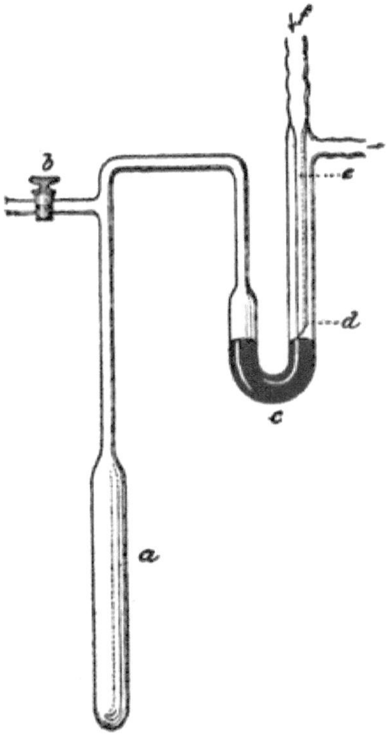

FIG. 27.—Bohr's Thermo-regulator.

hole is reached when the gas passing through it burns with a perfectly blue flame.

Rohrbeck's modification of the Lothar-Meyer regulator, is shown in Fig. 25. It is much more sensitive than the Reichert instrument, since it is regulated by mercury and ether, the vapor of which changes in tension comparatively strongly with slight changes of temperature. The gas enters at *a*, traverses the tube *d*, in which at the bottom an acute, sharp-angled triangular opening (*b*) is made, and passes out toward the burner at *c*. The mercury is indicated by line-shading, vapor of ether fills the chamber *g* [which is limited by a glass diaphragm *hi*, prolonged downward into a tube open at the bottom, which plunges well into the mercury]. When the apparatus is warmed, the mercury is forced by the expanding ether-vapor into the tube *hh*, say to the dotted line *i*, so as to partly close the triangular opening, which is then enlarged or diminished according as the mercury falls or rises. From the form of the opening, and its sharp angles, this adjustment occurs very uniformly. The first adjustment of the apparatus is effected by sliding the tube *d* up or down through the bored cork by which it is adapted to the larger tube. Fig. 25 show the regulator in its cheapest form. There is also a more improved form, in which the tube *d* is of steel, and adjustable by a fine screw (Fig. 24, *g*).

A regulator constructed by Bohr (Fig. 27), offers considerable advantages as compared with these commonly used models. The reservoir *a*, filled with air, is brought, with its stopcock *b* open, into the chamber that is to be kept under control. Shortly before the desired temperature is reached, the cock *b* is closed, after which every rise of temperature, however small, will cause an expansion of the air in *a* and a displacement of the column of mercury, which closes the opening *d*, allowing the gas to pass only through the reserve opening *e*. It must be seen that the inside of the reservoir *a* is not damp when the apparatus is put in use. Enough mercury is poured in through *f* to reach as high as *d*, so that this opening is entirely closed by a slight displacement of the mercury. It is advantageous to have the U-shaped tube constricted at *c*.

In addition to its great simplicity, the Bohr regulator is superior to those previously described in that the same instrument can be used at any desired temperature below the melting-point of the glass, and is adjusted for any temperature with extreme ease. It is also equally sensitive for all temperatures. The influence that considerable barometric changes exert on it (as well as on the Rohrbeck model) is readily compensated by opening the cock (*b*) for a moment. It further deserves mention, that by Bohr's regulator a constant mean temperature can be maintained in larger spaces, since the reservoir *a* may be given any desired form and size;

or a long lead tube with one end hermetically sealed (pinched or melted together) may be used as a reservoir, passing into various parts of the chamber.

A thermostat of this construction, with regulator, will generally be sufficient even for finer experiments, which demand a constant temperature during months. In most cases, however, we need only to have a chamber in which bacteria can be grown at something above 30° C., and for this a thermostat is perfectly good and useful, even if it varies a few degrees in the course of the day. If too much is not required, the regulator may be entirely dispensed with, and an ordinary gas, petroleum, spirit, or oil flame used. Of these four sources of heat, the first and last are unconditionally to be recommended. In case the apparatus is to be kept in our working room, petroleum is less satisfactory because of its ill-odor, independently of the greater danger from fire. Alcohol is better than [lard]-oil only in giving a flame free from smoke, while oil is cheaper and requires less care in its use.

It is naturally most convenient to use gas when this is at hand in the room. Since the gas flame as a rule needs to be only small, when a small thermostat standing in an inhabited room is to be kept at 30° to 40° C., a pointed yellow flame is always to be used (*cf.* p. 489). By a few days' experimenting, the proper height of the flame is ascertained, and when the temperature of the surrounding air does not vary too much this affords a good guide. A regular periodical change in the gas pressure, as at night, is easily compensated for by slightly raising or lowering the burner each day at the necessary time.

If one is unable or unwilling to use gas, [lard]-oil is to be recommended, especially if used as shown in Fig. 28. A large basin is filled to within a couple of centimetres of the top with water (*d*). Over this is poured carefully a layer of oil 2 cm. deep (c), upon which are placed one or several small floats (*f*) with wicks, such as are used on the continent for night-lamps. Such a floating burner gives an especially constant heat. As the flame cannot be changed in size, it is necessary to find the right distance between it and the thermostat; and since this must be kept as nearly constant as possible throughout the day, it is usual to employ a large bowl and a thin layer of oil, so that the wick shall not recede too far from the bottom of the brood-oven as the oil is used up. The small wicks are renewed morning and evening.

In Figure 28 is shown a still cheaper thermostat than that represented in Fig. 24. It has no regulator, no tube for showing the height of water, no faucet for removing this, and no attached covering of felt. But it can at any time be easily

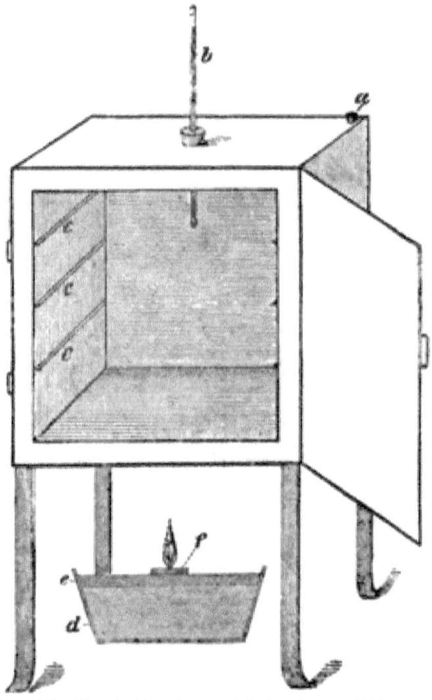

FIG. 28.—Simple Brood-oven Heated by an Oil-lamp

covered by rectangular pieces of felt or flat sheets of cotton-batting, of which four can be sewed together at their edges, while the fifth, over the door, is fastened only by its top to the upper piece, so that it can be lifted when the door is to be opened, while it is tied to the side pieces when the door is closed.

CHAPTER VII.

BACTERIOLOGICAL ANALYSIS OF FLUID, SOLID, AND GASEOUS SUBSTANCES; ESPECIALLY OF WATER, THE SOIL, AND AIR.

MICROSCOPIC investigation shows very incompletely what living germs are present in a preparation. Different appearing micro-organisms may represent different stages in the development of a single species, and bacteria which look quite identical may belong to entirely different species. Moreover, all germs of bacteria are by no means recognizable as such under the microscope; and the living are not certainly distinguishable from the dead. If the question is merely whether some specified virulent form occurs in a sample of earth or water, in case this form is pathogenic for animals, the most direct answer will be obtained by inoculation. Water is subcutaneously injected into a suitable animal. Earth is placed in a little pocket made in the subcutaneous tissue. Bacteria from the air are first collected in a sterilized filter of sand (*cf.* analysis of air), which is then introduced into such a pocket. But a satisfactory bacteriological analysis cannot to-day be made without recourse to cultures. It is of especial interest for the physician to know these methods, since they afford the most important introduction to hygienic investigation of water, earth, and air. Since the discovery of the cholera *spirillum* they have also been included among the best methods of clinical investigation. An exhaustive bacteriological investigation includes a determination of the number of bacteria present, as well as their identification; but for easily understood reasons, in both hygienic and clinical investigations, qualitative analysis is most frequently of entirely greater importance than quantitative.

A. Liquids.

(With especial reference to water analysis.)

All of the methods indicated in Chapter IV. may find application in the bacteriological analysis of fluids. Of these, we usually resort to dilution, and isolating the germs in nutrient gelatin, as the two principal methods.

The latter, first indicated by Koch, and developed by him to a high degree of completeness, is alone sufficient in by far the greater number of cases, and consequently will be treated here in detail. Koch's method consists in mingling the fluid containing germs with liquefied gelatin, which is then spread in a thin layer and allowed to solidify, being protected from atmospheric contamination. The mixing and spreading can obviously be effected in a great variety of receptacles, among which are four deserving especial attention.

a. Conical Flasks.—A medium-sized flask (Fig. 9, VII.) containing a thin layer of gelatin in the bottom is warmed just enough to melt the gelatin (over-heating may destroy the bacterian germs). By aid of a needle or pipette, a small quantity of the liquid to be analyzed is then added, with the greatest possible rapidity and care. The two are carefully mingled, but without shaking, which might cause the formation of bubbles. To avoid this, the flask is held upright, and several times carried around a large circle in a horizontal plane. The fluid is set in so active movement by centrifugal force that the commingling is perfect, without the introduction of bubbles into the gelatin, which is then allowed to solidify. This is one of the simplest and surest methods by which a Koch isolation-culture can be made. One disadvantage attends it, viz.: the isolated colonies are not accessible, in the flask, for microscopic investigation, and less accessible for examination by the naked eye or with a lens than in the following cases.

b. Test-tubes.—The mixture can also be effected in a test-tube, on the inside of which the gelatin is then allowed to harden in a thin layer (von Esmarch). For such "roll-cultures," wide tubes are best, with about 10 cc. of nutrient gelatin. Naturally, one may add the fluid as indicated in Chapter V., and subsequently melt the gelatin over a water bath, or this may be done first. The mingling of the two is effected as in *a*. The distribution of the gelatin over the wall of the tube is secured simultaneously with its hardening in the following manner: The projecting portion of the cotton plug is clipped or singed off, and the whole covered with a tight-fitting rubber cap to keep the cotton dry, and the tube laid in a dish of cold water, preferably ice-water. While the tube floats on the water, it is kept with one hand as nearly horizontal as possible, while with the other it is rotated until the gelatin has become solid in a uniform layer over the entire inside of the tube, which is then dried off and the rubber cap removed. Occasionally the cotton plug becomes covered with so thick a layer of gelatin as to prevent the access of air to the culture in an injurious manner. In this case, cotton and gelatin may both be pierced by a sterile platinum needle. The advantage of the test-tube over a flask is that it takes up less room, and its contents are somewhat more readily accessible for investigation.

c. Glass Trays.—It is convenient to use a pair of shallow trays (Fig. 12) about 20 cm. in diameter and 1.5 cm. deep, which are, of course, first to be sterilized, well wrapped in paper, from which they are removed only immediately before being used. Mingling the fluid which contains bacteria and the gelatin, is effected in a test-tube as in the last case. The cotton plug is removed and the mouth of the tube flamed [it is better to clip the cotton off close to the edge of the tube, push the plug down a little way, and then flame until the cotton browns a little. After cooling, the plug is easily removed with sterile forceps, and the gelatin poured.—W. T.], and as soon as it has cooled the contents are poured as rapidly as possible into the lower (smaller) tray, while the other, serving as a cover, is pushed a little to one side. Since the tube is never entirely emptied, but a small quantity of gelatin remains in it, it is best to label and preserve it as complementary to the tray culture.

A uniform layer of gelatin is seldom secured in these trays. The bottom is not plane, but as a rule a little elevated in the middle. But the consequent difference in thickness of different parts of the gelatin has its advantage, for when many germs are present in the fluid to be analyzed, and they are too closely crowded in the thick peripheral part of the layer, they are more isolated in the central thin portion.

The colonies in such tray-cultures are much more easily accessible for examination with the naked eye or lens than in *a* or *b*. Examination is frequently rendered somewhat difficult because the under side of the cover becomes clouded. In this case it may be replaced by a dry and sterile glass plate during the necessary inspection. [Non-liquefying species are readily examined by inverting the trays.] These cultures are also far better for microscopic examination than those in flasks or test-tubes, especially if smaller trays (5 or 6 cm. in diameter) are employed; but they are far from being as good in this respect as plate-cultures.

d. Glass Plates.—This method, the "plate method" of Koch, of which the three preceding are only modifications, dis-

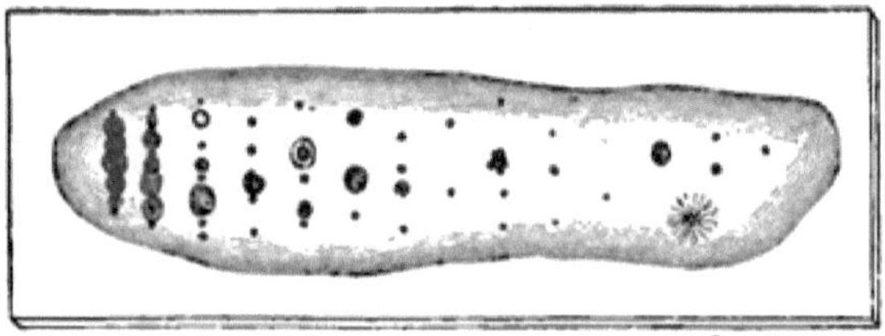

FIG. 29.—Slide Coated with Gelatin Crossed by Fourteen Inoculation Scratches bearing many Colonies. At *b*, a mould colony due to accidental contamination from the air.

tinguishes itself from these not only by a certain elegance, but especially in allowing a microscopic examination of the colonies anywhere on the gelatin, with high magnification. Koch's first plate-cultures were undertaken in the following manner: The gelatin was spread in an elongated drop on a common microscope slide, by the use of a pipette. When it had become solid, it was inoculated by a bent platinum needle (Fig. 21, *d*) dipped in the fluid containing germs, and then used to make a series of transverse scratches on the gelatin (Fig. 29). In the course of a few days colonies of the bacteria appear in the inoculation furrows,—closest and often entirely coalescent in those made first (*a*), fewer and well separated in the last. Assistance in recognizing contamination of such scratch-cultures from the air, is afforded by the fact that they almost always appear between the scratches, as is the case with the mould colony *b*, in Fig. 29. In two respects the plate method is inferior to the three preceding. From the nature of the inoculation, the plates are more exposed to atmospheric germs; and the distribution of the gelatin over the plates takes more time and care, since it must be kept from running over the edges, which is best avoided by the use of suitable cooling appliances which cause more rapid solidification.

The apparatus needed, and the method of starting a plate culture are, in detail, the following:

Rectangular glass plates 6.5 x 12 cm. are washed, polished with chamois skin and sterilized at 150° C., well wrapped in paper from which they are first removed immediately before use. [For convenience, the sheet-iron boxes used by the Germans to hold sterilized plates, are to be recommended.—W. T.]

The fluid containing bacteria is mingled with the gelatin as explained above (p. 56), in a cotton-plugged test-tube con-

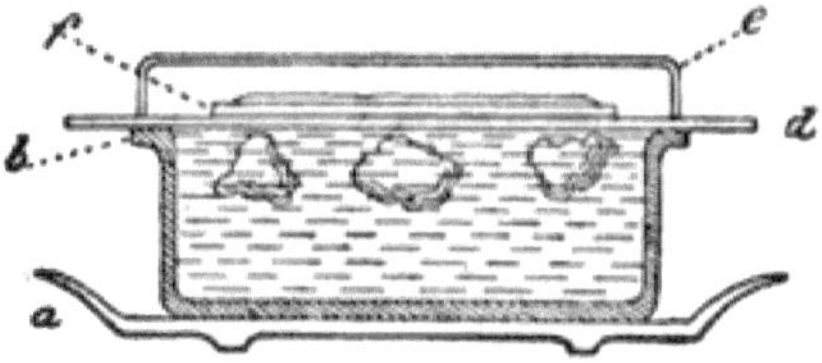

FIG. 30.—A Simple Cooling Apparatus for Plate-cultures.

taining about 5 cc. of peptonized gelatin, a quantity adapted to the size of plate recommended; especial care being taken to warm the gelatin only a little above the melting point.

A serviceable cooling apparatus is easily made as shown in Fig. 30. On an ordinary pie-plate (*a*) is set a preparation glass (*b*) with ground rim, several pieces of ice are placed in this and water is carefully poured in until the vessel is full, with a slightly convex surface, but not running over. A well-cleaned square plate of glass (*d*) is now let down on the surface of the water so as to wet it everywhere and thus avoid air-bubbles, a little water running over into the plate as this is done. Finally, a sterilized glass dish (*e*) is inverted on the plate to serve as a bell-glass.

One of the glass plates is unpacked and quickly laid under the bell-glass (Fig. 30, *f*), where it is rapidly cooled by the ice water. The plug is removed from the infected tube, the margin of which is quickly flamed [here, as above, p. 47, it is preferable to flame the tube before removing the cotton plug. —W. T.], after which the bell-glass is raised and the still fluid gelatin carefully poured over the glass plate, care being taken not to let it flow too near the margin. The bell-glass is replaced to protect the plate while the gelatin solidifies. The more exactly horizontal the plate stands, the better. If the gelatin runs in a given direction, because the plate is not level, it is possible to carefully lift the apparatus with both hands and check this by adjusting its position, until the gelatin is sufficiently hardened. It is naturally more convenient to level the plate beforehand by using a tripod and level (Fig. 31). If it solidifies too quickly, before spreading over a sufficiently large part of the glass, it may be spread by use of a sterile glass rod, warmed, but not too hot. The test-tube, in which

FIG. 31.—Leveling Tripod with Round Level (*b*) and Plate.

FIG. 32.—Bench of Sheet Zinc for Supporting a Plate-culture.

a few drops of gelatin always remain, is again plugged with the original cotton; labelled, rotated for several minutes and laid horizontally, so that this vestige solidifies in a thin layer, and kept as supplementary to the plate-culture, which, as soon as it has hardened, is labelled and quickly removed to the moist chamber.

For a single plate, a pair of shallow glass trays (Fig. 12) may be used as a moist chamber. But as a rule it is necessary to make several plate-cultures of the same fluid. In this case a pair of larger deep trays are used, or a chamber may be improvised. Before use, it is to be rinsed with 0.1 per cent sublimate solution, and a couple of sheets of filter paper, well wet with the same solution, are laid in the bottom. The plates are set in such a moist chamber, on small glass or metal benches which can be set over one another. Sheet-zinc benches of the form shown in Fig. 32 are very suitable because of their durability. They can easily be sterilized at 150° C. before use.

The examination of the plates by the naked eye is usually

made over either a white or black background (sheets of paper or tiles), as many differences in color between the colonies are only seen well in this way. It is also to be remembered that colonies of the same species and of equal age may present great differences in size, form, and color, according as they grow at the surface or immersed in the gelatin. Counting the colonies is best effected by the use of a lens of low power, and is much easier if the plate is laid on another divided into square centimetres. [When it can be afforded, one of the simple counting appliances used by the Germans in water-analysis, is to be strongly recommended.—W. T.]

The microscopic examination of the plate is not difficult, even with rather high powers. To be able to examine every part of the plate, a smaller size than the one recommended may be necessary, if the stage of the microscope is not quite large. Since it is impossible to calculate beforehand the number of germs added to the gelatin, in each case it is customary to prepare three plate-cultures at a time of three degrees of dilution. This is best done as follows: The gelatin is liquefied in three test-tubes, which are numbered 1, 2, and 3. After 1 has been infected by use of needle or pipette, and the added material well distributed, 2 is infected from it; and, in the same way, 3 from this. The gelatin is poured on the plates as above. In this way, it is highly probable that one of the three plates will contain a suitable number of colonies, whether these are wanted so numerous that the culture can be used for a tolerably reliable estimate of the number of germs present in the fluid (*cf.* water analysis, p. 64), or so few that a nearly absolute guarantee is given of the purity of each colony, and transfers may be made from them without difficulty.

It should be remembered, as advised above, to save the three test-tubes after distributing the gelatin in a thin layer by rolling them several times. Where the object is to get an exact determination of the number of germs in the original fluid, it is necessary to include those remaining in the tube, and, independently of this, it is always an advantage to have two isolation-cultures of each dilution.

Soyka obtains a large number of very small plate cultures by using pairs of trays which differ from those shown in Fig. 12 in being made of glass 1.5 mm. thick, with a bottom 50 square cm. in area, and walls only 3 to 5 mm. Seven round hollows are ground in the lower tray, like those in hollow-ground slides (Fig. 57). Before being used, they are slightly warmed, gelatin is placed in each of these depressions by using a pipette, one of the drops is quickly inoculated with a looped platinum needle, carefully stirred, and a transfer made from this to the second, etc., the upper tray or lid being put in place after all are inoculated. If the pipette is accurately graduated, and the quantity of fluid carried by the platinum loop previously determined by weighing it dry and again with its drop of water, it is claimed by Soyka that a sufficiently reliable determination may be made of the number of germs. This modification of the customary Koch method of plating out cultures appears to me excellent in its plan, but the form of the trays and the limited size of the hollows do not seem so good. For microscopic examination, I would recommend rectangular glass trays (*e. g.*, 12 x 5 cm.), the lower one divided by elevated ridges into eight quadrangular spaces, so that the surface of the several cultures should be larger.

In the preceding descriptions we have always assumed that the isolation was to be effected in either gelatin or agar.

1. *Gelatin.*—This is most easily used, since it may be melted and kept fluid at a relatively low temperature; yet the fact that many bacteria and moulds liquefy (peptonize) the gelatin is a disadvantage that often becomes very evident in such analyses.

2. *Agar-Agar.*—Most bacteria cause no liquefaction of this substance, which is nevertheless not so good as gelatin for plate-cultures, because of the difficulty of melting it and its somewhat slighter transparency, and its tendency not to adhere to the glass; still, it may be used well for analyses of this character if the following precautions are taken. When the agar has been melted in the test-tubes, which does not occur below 90° C., these are set aside to cool for a few moments and, before their contents begin to solidify, placed in a water-bath kept at 40° C., at which point the agar remains fluid, while the liquid containing bacteria can safely be added, without danger of killing any of the germs.

While it is usually necessary when gelatin is used to hasten its solidification by cold (or iced) water, this must commonly be retarded when agar is employed, by placing the plates over, or rolling the tubes in, lukewarm water, to prevent the agar from forming lumps and as a result hardening with an uneven surface. Since agar adheres so poorly to glass that it occasionally entirely separates from the plate and slips off, Fraenkel advises the use of a drop of sealing wax on each corner of the plate. [Where agar must be used in plating out bacteria, it is much better to do this in trays (supra, *c*), by which the curling of the agar is avoided.—W. T.]

3. *Agar-Gelatin*, prepared after Jensen's formula (supra, pp. 25, 26) combines some of the good qualities of both gelatin and agar. It is very conveniently used for plate-cultures, and, as a rule, is to be recommended in preference to either. The surest way of keeping it fluid during the process of inoculation, is to set the test-tubes in a water bath at 30° to 40° C.

4. *Serum* cannot be used in this manner, since it is coagulated only by prolonged heating to a degree likely to prove fatal to the germs. If it must be employed, one of two methods must be adopted: Scratch-cultures may be successively made with an infected bent platinum needle, in a large number of test-tubes, the colonies being so few in the last of the series as to be separate (*cf.* the original scratch-cultures on slides, p. 58); or the fluid to be used may be diluted with sterile water, and a small drop of the mixture allowed to flow about over the surface of the serum, in the hope that the germs will settle on it at a distance from one another. But it should be borne in mind that this is an exceptional method which does not give nearly so complete a separation of the germs as the careful distribution in melted gelatin.

Water Analysis.—The bacteriological analysis of drinking water is undertaken according to the rules given above. The following precautions are also to be taken: The water is collected in sterilized glass flasks, with cotton or glass stoppers. If it is taken from a faucet or hydrant, it is allowed to run for some minutes before any is gathered, and care must be taken to prevent the contamination of the stopper from any source (by placing it in a sterile glass box, Figs. 11 and 13, etc.). Samples are taken from accessible bodies of water in sterilized pipettes, and quickly distributed in sterile flasks. In case of deeper, less accessible supplies, a weighted sterilized flask is lowered and drawn up when full, its contents being at once divided between several smaller bottles. The analysis must be made as soon as possible after the sample is taken; even a few hours' delay at ordinary room temperature effects a great increase in the number of germs. If it is necessary to send samples of water to any considerable distance, or to let them stand for some time before they can be analyzed, they must be packed in ice.

The quantity of water to be added to each tube of gelatin, depends upon the number of germs present in it. When few are present, an entire cubic centimetre may be used. Occasionally $\frac{1}{10}$ cc. suffices. Water that contains germs in especial abundance must be diluted with occasionally several thousand times its bulk of sterilized water, and a drop may even then give a large number of colonies on the gelatin plate. Botton puts 10 colonies to the plate as a minimum, and 5,000 as a maximum, if good results are to be secured. With fewer than 10, accidental contaminations easily vitiate the results. If the plate contains too many, the task of counting them is more difficult, and some of them may not develop enough to become visible. It is best to keep the plates at 20° to 22° C., and to count the colonies on the third or fourth day.

B. Solid Substances.

The bacteriological analysis of solids is quickly described after the detailed account of fluid analysis. It may be carried out exactly like the latter, by first quickly dropping the substance into the melted gelatin, in which the germs are washed out and distributed as far as possible in the manner indicated for the analysis of water. If it is not certain that the germs can be sufficiently washed out even by continued agitation in the liquefied gelatin, this may be effected by violent and prolonged shaking in sterilized water or bouillon, which is afterward investigated according to the rules given for fluids. It is also possible, after such treatment, to add an equal quantity of melted 20-per-cent nutrient gelatin to the water, and, after careful mixing, to spread this out in one of the described methods.

Fraenkel recommends the first method for soil analyses, as the result of a large number of comparative experiment.[4] Because of the great difference between different samples of soil, as regards the amount of water present, he further advises measuring the samples in preference to weighing them, using for this purpose a sharp-edged spoon holding about $\frac{1}{50}$ cc. A level spoonful of dirt is poured into a test-tube containing melted gelatin, and broken up and distributed in this as thoroughly as possible by the use of a stout platinum needle, and by agitating the glass, after which the procedure is the same as for analysis of fluid. The same writer advises the use of von Esmarch's tube-method (*supra,* p. 56), and counts the colonies two days after preparing the tubes. But the common plate-method is equally applicable and occasionally even preferable, especially when the samples contain many liquefying forms, as is generally the case with the surface soil. A great deal of the difficulty from this cause may be avoided by the use of agar-gelatin. On the other hand, the tube-cultures admit of prolonged observation, and in this way of the detection of those very slowly developing forms which frequently occur in earth (*cf.* Chapter XII., on disinfection). Samples from the deeper soil, which, as the studies of Koch and Fraenkel show, always contain very few germs, can only be obtained with the certain exclusion of dirt from above by the use of a special boring apparatus (made by Muencke, of Berlin). Like water, soil must be investigated very soon after collection, since, especially in samples from some depth, there is a very rapid multiplication of the bacteria, which may become a thousand-fold more numerous than at first, in the course of a few days (Fraenkel). The reason for this increase, which cannot be prevented by the use of ice, is still unknown.

In his first soil analyses, Koch used the simple method of sprinkling the earth over the surface of gelatin. Naturally, no certain separation of the germs is possible by this means, but a good notion is obtained of the kinds of bacteria and moulds present in the sample. It is best for this purpose to pour the sterile nutrient gelatin upon plates or in trays, where it is allowed to harden. The earth is collected with the usual precautions in sterile test-tubes or flasks plugged with cotton. Immediately before use, the cotton is removed and a layer of filter paper quickly tied over the mouth of the glass. Holes are made in the paper with a flamed pin, and through these the dirt is dredged upon the gelatin, so that the fine particles do not lie too close together.

C. Air.

The micro-organisms which float in the atmosphere, and compose an important part of its so-called dust, have been investigated for a long time in many different ways, and with more zeal than those of water and the soil; on the one hand with reference to the theory of spontaneous generation, and on the other, because too much weight was laid upon the importance of the air as a carrier of nearly all contagious matters. But the time and energy bestowed are sadly out of proportion to the small resultant gain in knowledge of the life history and mode of transportation of pathogenic bacteria.

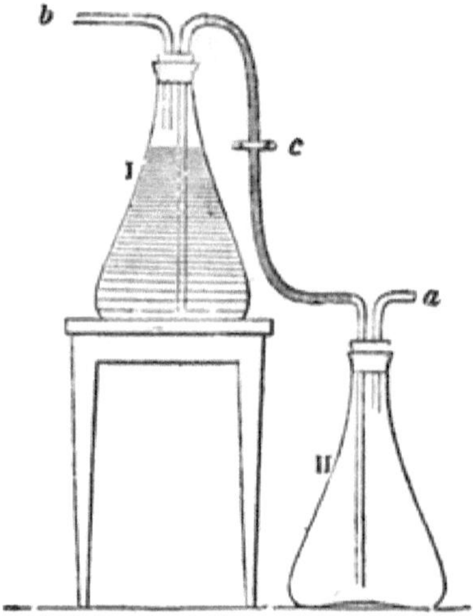

Fig. 33.—Aspirator Made from Two Flasks.

Two modes of collecting air and its organisms are in use:—the dust is allowed to settle by its own weight, or it is drawn into a suitable apparatus by using an aspirator.

Aspirators. — A simple aspirator, easy of transportation, is that shown in Fig. 33. Two large conical flasks (I. and II.) are furnished with rubber stoppers and glass tubes arranged (Fig. 33) as for wash bottles, and connected with a rubber tube in the manner shown. A pinch-cock is placed at *c*, and the upper flask filled with water. When the cock is opened and suction applied at *a* (or the upper flask tipped) till the siphon is filled, the water will flow without interruption from I. into the empty flask II., inducing a uniform and continuous suction of air into *b*, until I. is empty. By pouring the water back into I. (or reversing the relative position of the flasks) the suction can be recommenced. Knowing the volume of water used, and the time taken to empty one flask into the other, the amount of air sucked into the apparatus during a given time is also known. The rapidity with which the water flows from flask to flask can by regulated by the pinch-cock, or by the use of glass tubes of various calibre, inserted somewhere in the

FIG. 34.—Drip Aspirator (Filter Pump).

rubber tube. A number of such pieces of tubing are made ready, and the time required for each to permit the passage of a liter of water is experimentally determined and marked on it with a writing diamond, so that according to circumstances a larger or smaller tube is employed (Hesse).

Another form of aspirator is that shown above in Figure 7.

c. A drip-aspirator is shown in Figure 34 in which the constant stream from a faucet can be used as a means of suction. As the water falls in small portions from the faucet, through the tubes *a* and *b*, the little columns of fluid carry along between them larger and smaller columns of air, producing at *c* a powerful suction, which, however, is not so uniform as in the other two aspirators. [The German dealers supply a very simple aspirator essentially on this model, consisting of a T-shaped brass tube representing *abc*, joined by suitable corks and rubber tubing with a flexible lead tube some yards long, which, being non-collapsible, makes a very good connection with the other apparatus employed.—W. T.] Nor does the apparatus directly indicate the quantity of air sucked through, like the others, so that if this is to be determined it must be used in connection with a gas-meter.

I. *Collection of Dust for Direct Microscopical Examination.* — The dust which has accidentally accumulated in different places, often in a thick layer, can be simply collected, even if the layer is a slight one, or is easily swept on to a piece of paper with a feather. Small quantities of the dust collected in this manner are mingled with a drop of glycerin or other fluid on a slide and subjected to microscopic and micro-chemical examination in the usual manner.

If the dust is to be collected for a given time and place, this is best done by the use of a so-called aeroscope (Ponchet, 1859). Of the various models, only Schnauer's will be described here—a form which can easily be put together in any laboratory. It consists (Fig. 35) of a bell-glass (*a*) with ground base and short neck. A bottle with the bottom broken off and the lower end ground flat can be made to serve the same purpose. The neck is fitted with a rubber stopper provided with two holes, in one of which a short glass tube (*d*) is fitted, while a longer tube (*e*), bent in an arch above and drawn out to a point with an opening of about 1 mm. at the other end, passes through the second. The ground base of the bell-glass is smeared with vaseline and pressed air tight against a glass plate (*b*). Under the bell-glass is placed a slide (*c*) on which is a small drop of a sterilized solution of grape sugar 1 part, and glycerin 2 parts (Miquel). The tube *e* is then adjusted so that its point is about 1 mm. above the surface of the drop, and the tube *d* is connected with an aspirator. As soon as the suction begins, the air, and the dust suspended in it, flow through the curved tube directly against the drop of glycerin, in which the flow of air causes a slight depression, and where a large part of the dust is stopped. A part settles in the curved tube, which must, nevertheless, have this form in case the apparatus is to be used in the open air in rainy weather.

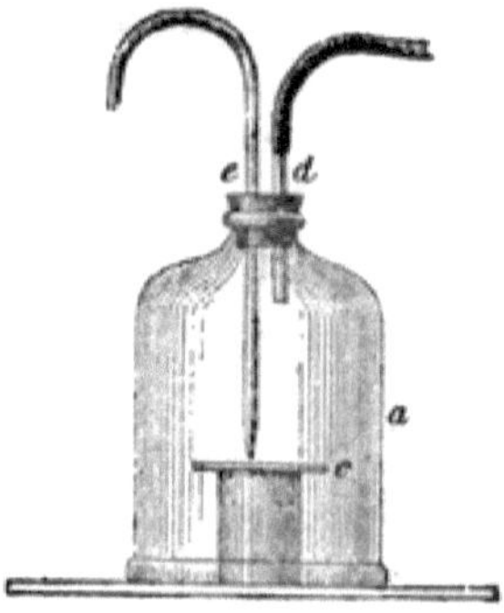

FIG. 35.—Schoenauer's Aeroscope.

When the aspirator has operated long enough, the slide is removed, and the dust uniformly distributed through the glycerin with a sterilized needle, a cover glass placed on it, and it is ready for microscopic and micro-chemical examination.

Preparations of this sort give a good notion of many of the inorganic and organic substances which float in the air in enormous quantities, *e.g.*, coal dust, sand, small crystals of various salts, woollen and cotton threads, hair, starch-grains, fragments of vegetable tissue, pollen, etc. Many mould spores are also recognizable, but the germs of bacteria are too small and too little characteristic to be identified in such preparations. The number and character of such germs present in the air at a given time can only be determined by sowing the dust in suitable culture-vessels.

II. *Collection of Dust for Cultures.*—Very pretty and useful preparations are obtained by simply letting the germs settle from the air upon the surface of gelatin. Several pairs of shallow glass trays (Fig. 12) are sterilized at 150°, each pair wrapped singly in paper. The nutrient gelatin (most commonly peptonized meat infusion with agar-gelatin) is carefully and rapidly poured in a shallow layer into the smaller tray, which is at once covered with the other (*cf.* p. 63). When the gelatin has solidified, the trays are wrapped up and taken to the place where they are to be used. Here they are unpacked, the lids (*i.e.*, the larger, upper trays) taken off, and the gelatin is left exposed to the air for a longer or shorter time, varying with the abundance of germs from a few minutes to an hour. After replacing the lids, the trays are set aside at room temperature, for observation, the germs beginning to show evidence of germination after a few days.

It is best to use the sterilized paper to wrap the lids in while the gelatin is uncovered, and, after they are replaced, to again wrap the pairs of trays in it for their return to the laboratory. Instead of pouring the gelatin directly into trays, it may, of course, be spread on a glass plate, which is then set within a pair of trays.

Simple as this method is, it gives, as has been said, very useful results. This may easily be shown by exposing a number of trays of gelatin simultaneously and for the same length of time in two different places. The differences in number and kinds of atmospheric germs in different localities is then very evident. It is to be recommended for the qualitative air analyses most frequently called for, but it does not give the number of germs contained in a given quantity of air. A number of methods and appliances for this quantitative analysis of atmospheric air have been indicated, but there is at present a difference of opinion as to which is to be preferred. I have not sufficient experience in this direction to pronounce very authoritatively on the question, but I have no doubt the air-filters using (insoluble or soluble) powders are to be given the preference. In view of this difference of opinion it must be a satisfaction to know that at present the exact determination of the number of germs in the air is not of very great importance. I shall, therefore, merely describe briefly the choice of apparatus representing the various systems, viz., automatic suction balloons, settling apparatus, such in which the air is allowed to bubble through fluid or gelatinized contents, and filters, which may be insoluble or soluble.

The automatic balloon was first used by Pasteur in 1860. As shown in Fig. 36, it is at once culture vessel, aspirator, and gasometer. It consists of a globular flask with a long neck drawn out into a capillary tube. The tube is hermetically sealed, and the flask sterilized at 150° C. When cool, the point is broken off, the air within the balloon is warmed a little over a burner, and the tip dipped into the sterilized culture-fluid. As it cools, a few drops of fluid pass into the flask, this is carefully heated to the boiling point over the burner, and, while the steam is escaping from the opening, this is again dipped into the fluid, a greater quantity of which is sucked in. When the flask is about half full, the point is removed from the fluid, and the contents carefully heated over the flame and allowed to boil gently for a couple of minutes. While the steam is escaping with a whistling sound from the end of the tube, the flame is removed, and the point is at once carefully fused. The apparatus is now ready for use, for which it is held as far from the experimenter as possible, in the direction from which the wind comes, so that germs shall not be blown from one's body toward it. The point is opened by a pair of pincers or shears that have been flamed, and the air rushes in with a whistle and fills the receptacle. The opening is once more sealed, the flask shaken so as to wash all germs from the walls into the fluid, and the development of micro-organisms in the fluid is awaited. [But this gives no quantitative determination, and a satisfactory investigation of even the aerobic forms present in the fluid is only possible when some of the fluid after thorough shaking is at once taken, with the usual precautions, as a basis for some form of plate-culture. —W. T.]

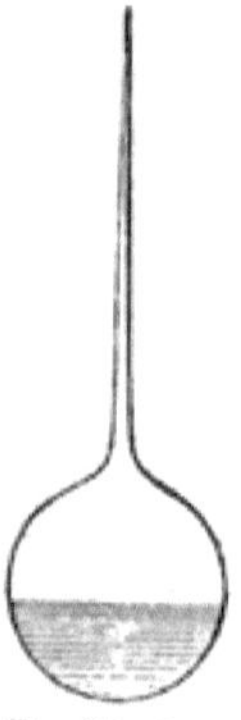

FIG. 36.—Automatic Suction-Bulb for Collecting Air-germs.

Hesse's aeroscope (Fig. 37) consists of a glass tube 50 to 70 cm. long and 4 cm. in diameter, one end of which is somewhat flaring and closed by a perforated rubber stopper (*d*), through which passes a short glass tube 10 mm. in diameter, plugged with cotton (*b*). The flange at the other end is somewhat more prominent; over it are tied two thin rubber caps, one outside the other, the inner of which is pierced with a round hole about 10 mm. in diameter, as shown in the section in Figure 37. The tube is used with nutrient gelatin, the surface of which must not reach above the lower side of the small glass tube or the hole in the inner cap, when the apparatus lies flat.

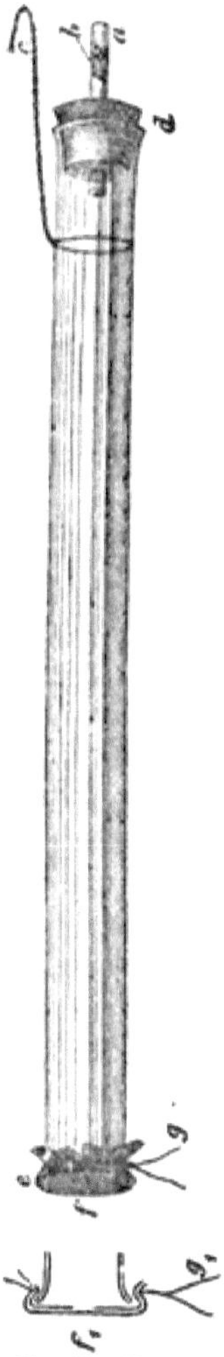

FIG. 37.—Hesse's Aeroscope.

The tube, stopper, and caps are cleansed in 1.0 per cent sublimate solution, and finally rinsed with boiled water. The inner cap is tied fast with thread, without being much stretched. If it is found to be water-tight, by half filling the tube with water, and holding this end downward, the round hole (*f* and *f'*) is clipped in its centre with sharp scissors, and the outer cap is tied over it, but in a more tense condition. [Rubber caps, fitting quite well and easily applied, are also supplied with the apparatus by dealers.—W. T.] Its power to hold water is then again tested. The stopper with its plugged tube in place is now inserted in the other end, and the apparatus, still containing water, is hung in the steam cylinder, which needs to be lengthened for this purpose (Fig. 4, C). A piece of wire (Fig. 37, *c*) twisted around the upper end is used for suspending the tube to the lid by the means shown in Figure 4 A. After exposure to streaming steam for a quarter of an hour, it is removed from the sterilizing cylinder, and when it is somewhat cooled the water is poured out and replaced with all precautions by melted sterile nutrient gelatin, poured from a pipette or wash-bottle. The stopper being again in place, it is once more hung in the steam-cylinder for ten minutes' sterilization at 100°, after which it is placed horizontally while the gelatin cools [or rotated until the gelatin is nearly hard, when it is allowed to lie flat so that one part of the circumference is more thickly coated than the rest. —W. T.] When wanted for use, it is set upon a suitable support, where the air is to be examined. Hesse uses a folding tripod such as is used by photographers, which ends in a flat surface on which the tube can be fastened, the two flasks of the aspirator (Fig. 33) being hung at different heights on the legs. Simple supports, answering every purpose, are also easily improvised. When the apparatus is set up, it is connected with the aspirator, the surface of which is carefully washed with 0.1 per cent sublimate, and the aspirator set in operation, the siphon being sucked full beforehand, to prevent a sudden irregular suction when the pinch-cock is opened. The quantity of air which Hesse advises to be passed through the apparatus is 1 to 5 litres for inhabited rooms, and 10 to 20 litres out of doors, but in certain cases much greater or smaller quantities may be advisable. He recommends the aspiration of a litre of air through the apparatus in one to three minutes for inhabited rooms, and the same quantity in three to four minutes for the open air, as a proper rapidity. After the air has been sucked through, the apparatus is closed by replacing the rubber membrane (which has been rinsed in sublimate), and set aside at room temperature. In the course of the next eight or ten days, colonies of moulds and bacteria grow on the surface of the gelatin, from which they can be obtained for examination and the inoculation of cultures by using a long glass rod bearing a platinum needle. If the suction has been effected sufficiently uniformly and with the right rapidity, so that all germs have been allowed to settle from the air passed through, the colonies should be most numerous near the end through which the air enters, gradually decreasing in number toward the other end, the last third or quarter of the gelatin surface being entirely free from colonies.

Bubbling Apparatus.—Of the different forms with fluid contents, only one, long employed by Miquel, is described here. This is a flask closed above with a cap plugged with cotton (*a*), like a Pasteur flask (*cf.* Fig. 10). The neck is prolonged as a fine tube to the bottom of the flask. The slightly constricted side tube *b*, is plugged at two points with cotton (not shown in the figure), and is connected with some form of aspirator by the rubber tube *e*. A capillary tube hermetically sealed at *d*, is connected by the same means with the other lateral tube (*c*), which is slightly bent downward. 30 to 40 cm. of distilled water is poured into the flask. Sterilization is effected as usual. The aspirator is connected with *e*, the plugged cap *a* is flamed and removed, the observer withdraws from the apparatus, and the suction of air into it is allowed to go on, after which the cap is again flamed and replaced. By blowing through *e* the water is made to rise and fall through the neck of the flask ten times, to wash off adhering germs. The point *d* is flamed and broken off and by tipping the flask and blowing through *e* the contents are divided among 30 or

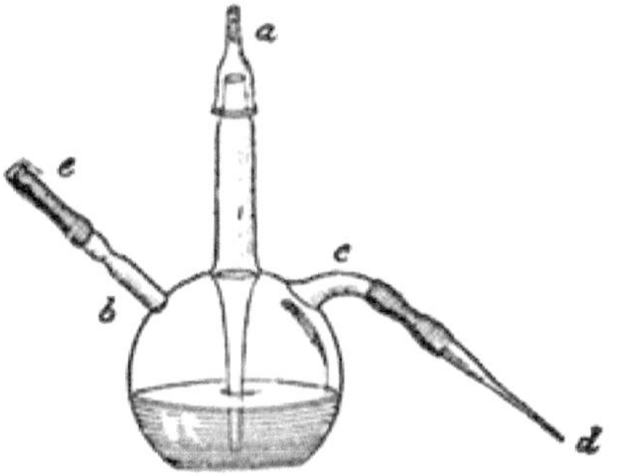

FIG. 38.—Miquel's Aeroscope.

40 culture flasks containing bouillon. Finally, the outermost of the two cotton plugs in the tube *b* is removed and the inner one pushed into the flask with a sterilized platinum wire, after 25 cc. of sterile bouillon has been poured into the flask. The flask, and the 30 culture-vessels are set aside at 30° C., and observed for at least a month. When a small number of the vessels show growth, the total of the germs collected by the aeroscope can be approximately calculated (*cf.* p. 41, method of dilution).

Tryde, Hueppe, and v. Sehlen formerly caused air to bubble through melted nutrient gelatin, in their aeroscopic analyses. Quite recently, Straus and Wurtz have used the same method, and for this purpose produced the glass apparatus shown in Figure 39. The top and bottom measure 15 mm. in diameter, while the central portion (*A*) is a wide cylindrical receptacle bearing the slightly constricted lateral tube (*D*), with two cotton plugs (*f* and *g*). The tube *B* is drawn out to a fine point below, and closed above with a cotton plug (*e*). At *c* it is widened, and ground to fit the mouth of *A* air-tight. The apparatus is sterilized at 150° C., after which 10 cc. of 10 per cent gelatin (or preferably agar-gelatin) are poured into it and a drop of sterilized olive-oil added, after which all is sterilized in the steam cylinder. The addition of oil prevents the gelatin from foaming, even when air is passed through it very rapidly, *e.g.*, 50 litres in ten minutes. The gelatin is melted, *D* is joined by rubber-tubing with an aspirator, and the plug *e* is removed. While air is passing through, the gelatin is kept fluid by the warmth of the hand, or, if agar-gelatin is used, a water-bath is employed. Afterward, the plug *e* is replaced, and, by blowing into *D* the fluid is forced into *B* several times to wash out any adhering germs. The plug *f* is removed so as to allow *g* to be pushed into the gelatin with a sterile platinum wire, after which it is replaced and *g* is gently shaken about in the gelatin, which is then allowed to solidify as for an Esmarch tube-culture, or poured out in glass trays as described on pages 496, 497. The observation of the development of the colonies is rendered not a little difficult by the fact that, while air is bubbling through, the oil becomes finely emulsified in the gelatin, rendering it cloudy.

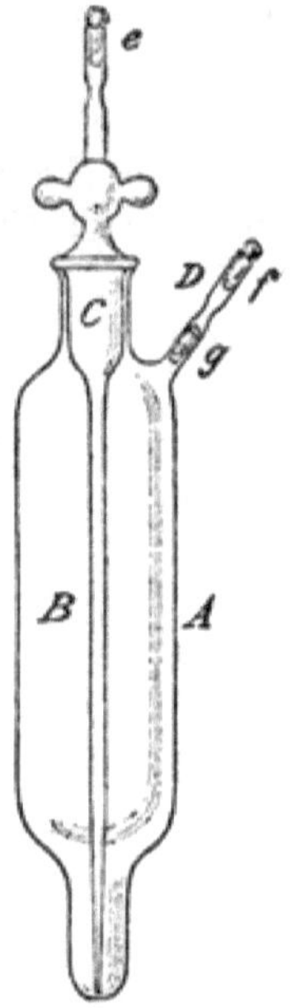

FIG. 39.—Aeroscope of Straus and Wurtz.

Insoluble Powder Filters.—As the result of a series of very carefully conducted experiments, Petri has recommended sand as a filtration medium in preference to glass-wool and asbestos, which have been used by others (Miquel, Moreau, Freudenreich, and Frankland). The details of his method are as follows: sand is passed through a sieve with meshes 0.5 mm. wide, and what passes this is sifted through a second with meshes 0.25 mm. wide, the portion that remains in this consisting of grains of the right size. This sand is heated to redness in an iron crucible (half or three-quarters of an hour being required for 100 cc.), while it is stirred with a glass rod. While warm, it is filled into sterile test-tubes plugged with cotton. Two sand-filters (s_1 and s_2) 3 cm. long are brought into a glass tube 1.5 to 1.8 cm. wide and 9 cm. long, as shown in Figure 40. The sand is held in place by small caps (n_1 to n_4) of fine brass gauze (40 meshes to the linear centimetre), which are easily made over the end of a glass tube or metallic rod. This separation of the sand in two parts enables the experimenter to sow the second sand plug separately, as a control filter. The filters being 1 cm. from each end of the tube, it is plugged at *f* and *g* with cotton, and sterilized at 150° C.

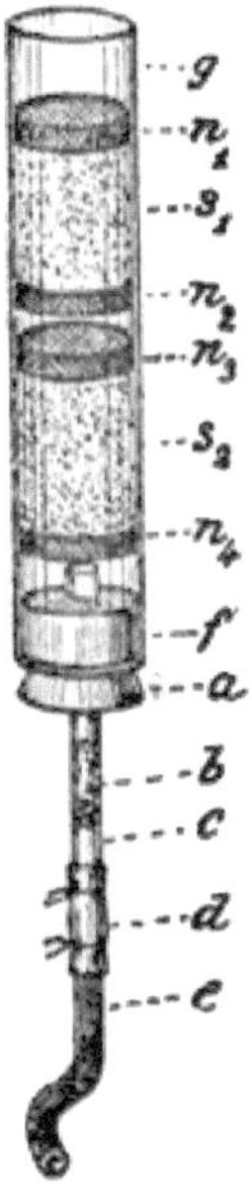

FIG. 40.—Petri's Sand-Filter.

Before use, the cotton plugs are removed and laid aside in a sterilized glass box. The filter-tube is joined with an aspirator by means of a rubber stopper (*a*) sterilized in sublimate, pierced for a cotton-plugged sterile glass tube *c*, which is connected by a short piece of rubber tubing with a lead tube (*e*) 0.5 cm. wide. The filter tube being arranged vertically, the cotton plug is removed from *g* and the suction can begin. Petri uses as an aspirator a hand air-pump, each stroke of which draws a litre of air through the filter. When, as in his experiments, 100 litres pass through the filter in 10 to 20 minutes, all germs are stopped by the first filter. The material is sown in pairs of trays (Fig. 12) 9 cm. in diameter, the sand of the first filter being divided between three, and of the second, between two such trays. The sand is first poured into the trays, and melted gelatin is poured over it from test-tubes (*cf.* p. 57). After the sand is completely wet by the gelatin, so that all air bubbles have been driven out, it is equally distributed by a series of short but forcible horizontal movements, and the gelatin is allowed to harden, when the three pairs of trays are set away one over the other at the temperature of the room.

Soluble Filters.—Miquel, the most experienced living student of the air, has very recently recommended [5] the employment of soluble filters, an ingenious method suggested by Pasteur over twenty-five years ago. Of the various powders which may be selected (cane-sugar, table-salt, sodium phosphate, magnesium sulphate, etc.), Miquel recommends cane-sugar, which demands no special preparation, may be sterilized at 150° C., without losing its high degree of solubility or becoming more hygroscopic, or without having any injurious effect on the bacteria. In very foggy weather it may become impossible to work with either insoluble or soluble filters. In this case a bubbling apparatus may occasionally have to be used. The sugar (if loaf sugar is used) is powdered in a mortar and shaken through two metal sieves, as described for sand, so that the size of the grains used is about half a millimetre. This sugar is placed in an aeroscope (Fig. 41, *A*) consisting of a glass tube about 20 cm. long and 5 mm. in diameter, narrowed at *e*, and, according to the suggestion of Freudenreich, provided at *b* with a ground cap plugged with cotton, as for the Pasteur-Chamberland flasks. Two plugs of (glass-wool or) cotton are inserted at *d* and *f*, and the whole is sterilized at 150° C. The cap is removed, and enough well-dried sugar (1 to 2 gm.) is poured into the tube to fill it for a length of 8 to 10 cm., after which all is again sterilized at 150° C. When it is to be used, by a series of light percussions the sugar is packed against the plug *d*, and the tube held nearly vertically (tipped toward the wind), so that during suction the sugar may not fall away from the walls, thus rendering it possible for the air to pass unfiltered at any point.

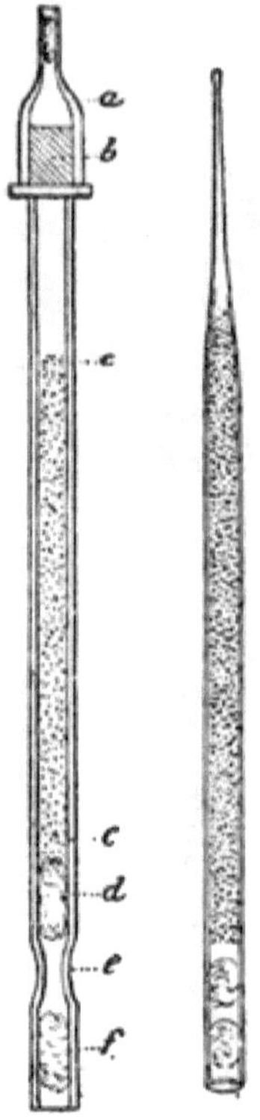

A. B.

Fig. 41.—*A*, Newer, *B*, Older Model of Miquel's Soluble Powder Filter.

The air is drawn through the filter with different rapidity and for a varying length of time, according to circumstances. The average number of germs for the day at a given point is obtained by slowly drawing the air through a filter for 12 or 24 hours. To learn the number at a given time, it is necessary to draw the largest possible quantity of air through the sugar in the shortest time possible.

After the aspiration, the cotton or glass-wool plug is removed and the sugar is poured into sterilized water. When many germs are expected, a large quantity of water (500 to

1,000 cc.) is used; otherwise a smaller quatity (50 to 100 cc.) is taken.

In case the sugar does not run out of the tube easily, it must be pushed out by the aid of the cotton plug *f*, and a sterile glass rod introduced at *f*, and the tube afterward rinsed out with sterile water. The plug *d* is also to be sown separately as a test of the efficiency of the filter, which should remain free from germs after the air has been sucked through. To simplify these manipulations, I advise the use of tubes that are not constricted at *e*, and to proceed as follows: The following articles are to be made ready beforehand: 1, an Erlenmeyer flask, filled with sterile water; 2, a short test-tube, with a few cc. of water; 3, a test-tube with nutrient gelatin; 4, a sterile glass box (Fig. 11 or 13); 5, a small rubber tube. The plug *f* is removed with forceps, and laid in the glass box, after which *d* is similarly removed and cast quickly into the gelatin, *f* being immediately replaced. The cap *a* is then taken off, and the sugar poured out into the Erlenmeyer flask, the rubber tube is slipped over the unground end of the glass (*f*). the ground end being dipped into the sterile water of the short test-tube, when all remaining germs and sugar are rinsed out by alternately sucking water into the tube and expelling it; finally this water is poured into an Erlenmeyer flask. The filter tubes which Miquel used in his earlier investigations were drawn out to a capillary point at the end through which air was admitted (Fig. 41 *B*, where the constriction between the two cotton plugs is also left out), and were opened and closed by breaking off and remelting the point. The use of the ground cap *a* renders it far easier to empty the apparatus, but adds materially to its expense; there is, however, no reason why simple glass tubes, closed by my rubber-tubing cap (*cf.* Fig. 9, III.-VI.) should not be used. The gelatin in which the plug *d* was placed should remain sterile, all germs being found in the Erlenmeyer flasks. When the sugar is entirely dissolved and the solution well shaken, the number of germs is determined exactly as described for water analysis, preferably using agar-gelatin, so as to avoid as far as possible the disturbing liquefaction of the gelatin.

A great advantage of this method is that only a fraction of the sugar solution and the germs it contains need be used for quantitative analysis, while the greater part can be used for various qualitative investigations (*e.g.*, a search for anaerobic forms), which it was not possible to carry out in the earlier methods in which all of the collected germs had to be used for the quantitative analyses.

Of the aeroscopes and methods of analysis here described, the first two have really only historical interest. The automatic suction balloon was the instrument with which Pasteur long since outlined the bacteriological topography of the atmosphere, which Miquel has since persistently worked upon. Hesse's apparatus marks the first attempt to employ the gelatin cultures of Koch in quantitative analysis of the air. Of the other four, doubtless the solid filters will deserve preference. They are easily transported, and can be "charged" with bacteria at any temperature and at places far from the laboratory, after which they can be kept (at least for several days) until cultures can be started under favorable conditions. They also accomplish what is possible toward collecting in a relatively short time all germs from the air of a very large space. Of these solid filters, the soluble appear to me far superior to the insoluble.

In this chapter, we have as good as exclusively concerned ourselves with the application of Koch's methods of isolation in gelatin to bacteriological analysis. It must, also, be said unqualifiedly to be the chief method of such analysis. Starting plate-cultures in one or other of the described forms is the means first to be employed when bacteria are to be found and separately cultivated from a given substance, and they will in most cases lead to the desired result if sufficient attention is given to varying the culture-material, temperature, time of observation, number of germs, etc. The later the colonies are counted, the more probability will there be that all germs have developed into visible colonies. It is not so much the danger of atmospheric contamination which leads to an earlier examination than might be wished, as the presence of rapidly growing, liquefying bacteria and moulds, which quickly spread, so as to cover large portions of the gelatin. But it is self-evident that it is not a universal method in the sense of rendering every other method superfluous. On the contrary, we may often do well to use the various other methods described in Chapter IV., either alone or as complementary; but no general rules can be given as to this, so that one must choose his plan for each respective case.

Finally, a most important point must be noted with reference to all the methods thus far described, viz.: They are calculated for only aerobic bacteria. An exhaustive analysis of an unknown mixture of bacteria, includes also a determination of the pronouncedly anaerobic forms, which play so important a part in the economy of nature, as well as in the causation of disease. In what precedes, these have been left entirely out of consideration, being reserved for special and detailed discussion in the next chapter.

CHAPTER VIII.

CULTURE OF ANAEROBIC BACTERIA.

In 1861, Pasteur made the surprising discovery that organisms exist which can live, nourish themselves, and propagate, without access to free oxygen. His doctrine of anaerobiosis ("la vie sans air") was received with doubt and mistrust; but further investigation not only corroborated his observation, but from this as a starting point, Pasteur succeeded in leading the physiology of respiration into new channels, bringing us nearer to an understanding of many important processes of fermentation. Anaerobiosis showed itself to be a very significant and wide-spread phenomenon, and a closer investigation of the need for oxygen of different bacteria (as well as yeasts and moulds) revealed an extreme variability in this respect. Some bacteria demand a very abundant supply of oxygen, while others are affected by it as a poison, which not only hinders their development, but quickly destroys their life. Between these extremes—the marked aerobiotic and anaerobiotic forms—every degree of transition is to be found. Some forms seem to thrive about equally well with or without free oxygen; some grow freely exposed to oxygen of a low tension, others when this is greater—a circumstance which occasionally finds expression in a very striking manner in the form and appearance of gelatin cultures. While the growth around the puncture made in the gelatin occurs for many species very abundantly at and immediately below the surface, and diminishes toward the bottom (Fig. 42, *c*), other cultures may be found (Fig. 42, *a*) which show an extremely slight growth near the surface, while the colonies of bacteria increase uniformly in size the further they are removed from the atmosphere. The former is consequently aerobic, the latter anaerobic. Beside these more or less conical cultures, with the point directed downward or upward, others occur as capillary or thick cylinders along the needle puncture (Fig. 42, *b*). The oxygen of the air has neither helped nor hindered the growth of these.

The discovery of anaerobiosis has necessitated the introduction of new apparatus and methods in the cultivation of bacteria. These must be treated now in detail, especially since we already know several pathogenic forms which are markedly anaerobic (malignant œdema, charbon symptomatique). Very different means have been employed for keeping the oxygen of the air from culture vessels and material,

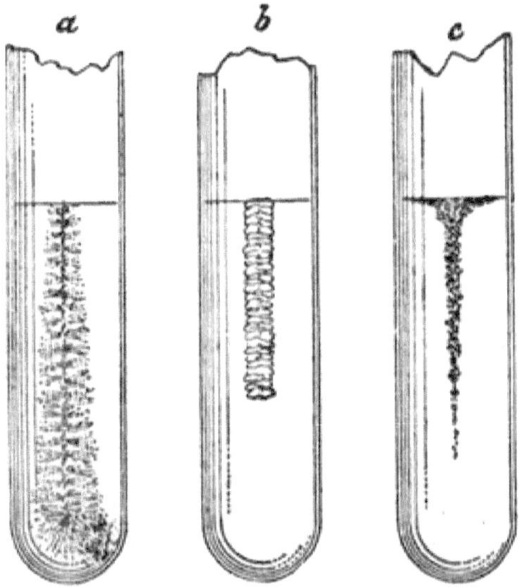

FIG. 42.—Test-tube Cultures, *a*, *B. murisepticus*, in agar-gelatin; *b*, a colorless yeast in raisin-gelatin; *c*, *B. Anthracis*, in agar-gelatin. The three cultures are of like age, and grown under exactly the same conditions.

or for removing all absorbed oxygen from the latter. Sometimes the effort has been limited to covering the culture-medium with solid substances (glass, mica, agar-agar), fluid (oil), or gas (carbonic acid, hydrogen). If good results are desired, the free oxygen held by the culture-medium must also be removed by boiling, the use of the air-pump, the passage of a gas which is not a supporter of combustion, but at the same time not poisonous (hydrogen), or by the aid of aerobic bacteria or chemicals with an affinity for oxygen.

Examples of most of these methods are described below,

though I shall confine myself to a discussion of a limited number of kinds of apparatus which have received the recommendation of expert investigators. In accordance with the plan of the book, the principal weight will then be laid on such methods as demand the most readily obtainable and cheapest material. Those who wish to go further into the technology of this branch of the subject are referred to the two chief treatises on the subject, by Roux [6] and Siborius.[7]

Even by the best of the methods described, it is not possible to remove the last trace of oxygen from the culture-vessels, but they have shown themselves sufficiently free from oxygen under all circumstances for the cultivation of the most pronouncedly anaerobic forms. As an easily applied test for freedom from oxygen, sufficiently accurate for our purpose, indigotin is to be recommended. Enough of a very dilute solution of the sulphate (1 gm. indigotin to a litre of water) is added to the culture-medium to give it a distinct blue color, a small quantity of grape sugar is added, and the reaction rendered distinctly alkaline by the addition of potash. Three drops of a 10-per-cent solution of potash are added to 10 cc. of nutrient gelatin or agar, which in addition to the usual substances also contains 1 per cent of grape sugar (Liborius). When (by means of boiling, the passage of hydrogen, or the use of the air-pump) the oxygen absorbed by the blue culture substance is removed, the latter becomes colorless, since the indigo-blue is reduced by the grape sugar to indigo-white. A return of the blue color shows that oxygen is not certainly enough excluded.

Trustworthy and adequate apparatus for the cultivation of anaerobic forms can only be prepared when a hydrogen developer or air-pump can be used. For the former, Kipp's apparatus is commonly used, though, as Jörgensen has shown, any one can fit up a sufficiently good generator at a cost of about half a dollar, by using a common student-lamp chimney, fitted by a loose cork (Fig. 43, *a*) into a cylindrical glass which contains a mixture of one part by weight of sulphuric acid and eight parts by weight of water (prepared with care by adding the acid to the water while it is being stirred,—never by adding water to the acid), and a couple of drops of platinum chloride solution. A perforated lead plate covered with muslin is placed at *b*, to support a quantity of granulated zinc in pieces about as large as peas. The stopper *c* must be of rubber, and fit perfectly. Hydrogen is evolved as soon as the zinc is brought in contact with the acid; but if the rubber tube at *d* is closed by a pinch-cock, the hydrogen drives the acid down to below *b*, and its formation stops, to recommence when the pinch-cock is again loosened. On the way to the culture-vessels, the gas is passed through an alkaline solution of pyrogallic acid (*e*) (1 part of a 25-per-cent aqueous solution of pyrogallic acid is thoroughly mingled, in a perfectly full bottle, with a 60-per-cent solution of potash) to remove a trace of oxygen which may be present.

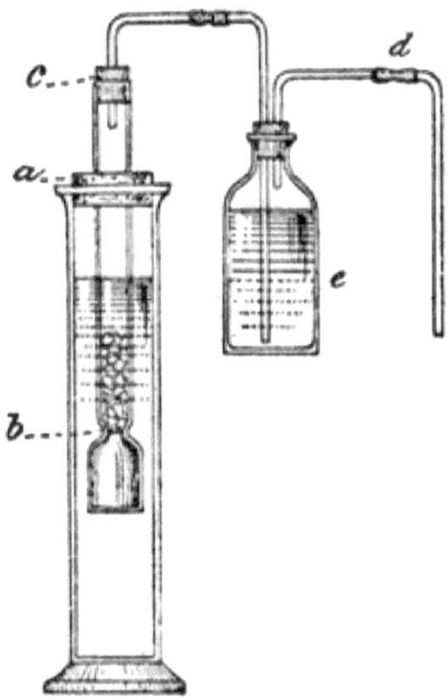

FIG. 43.—Joergensen's Hydrogen-generator.

The air-pump belongs to the class of expensive instruments which it is not proposed to discuss here in detail. It is, indeed, a great comfort to have the use of a mercurial air-pump when the cultivation of anaerobic forms is to be undertaken, but it is not a necessity. A comparatively cheap and good model is the Sprengel pump, as modified by Hüfner. One of the simple water air-pumps cannot be relied on for the removal of oxygen as completely as is sometimes necessary; but by also employing hydrogen as indicated by Roux (*infra*, p. 87), it may be effected with sufficient completeness.

The vessels and methods for cultivating anaerobic forms will be considered under two principal classes, according as they are to be used for isolating and obtaining pure cultures of anaerobic forms from a mixture of bacteria, or merely for propagating a species already pure, either in culture or in a diseased animal.

I. Isolation from a Mixture of Species.

Pasteur succeeded in getting the first pure cultures of anaerobic forms in fluid media, by the method of isolation based on physiological differences of species, using carbonic acid or exhausting the air. More recently, the attempt has been especially made to apply Koch's methods with gelatinized media, even to the analysis of mixtures containing anaerobic bacteria.

A. Simpler (and more incomplete) means, which do not require the use of either air-pump or hydrogen generator.

1. *Plate Cultures under Mica* (Koch).—Koch made the attempt to effect the plate-culture of anaerobic forms by covering the gelatin, while still soft, with a sterile sheet of mica. This must usually be as thin as a sheet of paper, without flaws, and sterilized by flaming immediately before use (or at 150° C., wrapped in paper), and care is to be taken not to allow air-bubbles to slip under it when it is laid on the gelatin. To secure a still surer exclusion of oxygen, the margin of the mica can be covered with melted paraffin (Fraenkel), which quickly hardens, forming a solid border. In this way, the usual method of working is preserved, as well as an easy accessibility for microscopic examination.

It is easy to convince one's self that the mica plate really hinders the access of oxygen to germs beneath it, to no small degree. If a decidedly aerobic non-liquefying form has been plated out in this manner, the colonies may be seen to appear in numbers in the uncovered part of the gelatin, while beneath the mica they occur sparingly near the margin, and in the beginning may be entirely wanting in the centre. Yet the method is frequently unsatisfactory, because the exclusion of oxygen is not sufficiently complete and lasting, and it is inferior to those described below, especially that recommended by Liborius.

2. *Isolation in a Very Deep Solid Medium* (Liborius).—A small quantity of the mixture to be analyzed is sown in a test-tube filled to a depth of 10 to 20 cm. with nutrient gelatin or agar, which has been melted and cooled to the lowest temperature at which it will remain fluid. The material is distributed as uniformly as possible by aid of a thin sterile glass rod carried carefully through a rotary as well as vertical movement in the gelatin. If the distribution is successful, and the number of germs right, very pretty results may be obtained, the number of aerobic forms decreasing in size and number from the top, while on the other hand the anaerobic forms

grow only toward the bottom, and are always wanting from the uppermost part of the gelatin. In either case, the distribution of the colonies always gives a good basis for deciding as to their need of oxygen.

The colonies of bacteria are naturally far less easily accessible for examination in the depth of such a test-tube than in a plate-culture. When there are few of them, and each colony is large and liquefies the gelatin, material for new cultures or for examination is easily obtained by thrusting a sterile capillary tube, open below but sealed at top, through the gelatin until the colony is reached, when the top is broken off. In other cases, it may be necessary to break the tube and to cut the gelatin up in a sterile tray so as to render the colonies accessible to the needle. If agar is used, this can usually be blown out of the tube by means of a Pasteur pipette (p. 44) passed down at one side. The microscopic examination of the colonies under moderately high powers can likewise often be conveniently made only after removing the contents from the glass and cutting it in thin slices.

FIG. 44.—Isolation-Culture of Anaerobic Bacteria in a Deep layer of Gelatin.—(Liborius.)

B. Methods in which hydrogen is used to replace the air.

3. *Isolation in Gelatinized Media.*—Roux recommends a tube similar in all essentials to that shown below in Figure 56, but of considerably larger size. After hydrogen has been passed for a sufficiently long time through the liquefied and inoculated gelatin, this is allowed to solidify while the tube is horizontal. When the colonies are developed, the tube is cut lengthwise with a diamond into two boat-shaped halves, the one containing the gelatin being then easily accessible for observation.

Fraenkel uses the following method: The test-tubes, rather thick-walled, of large diameter, and without spreading rims, are plugged with cotton sterilized in the usual way, and inoculated in three degrees of dilution as described on p. 500, after which the plugs are replaced as rapidly as possible, with sterilized fingers, by closely fitting rubber stoppers with double perforation, containing two glass tubes bent at right angles above the stopper. One of these reaches the bottom of the test-tube, while the other ends immediately below the stopper. The horizontal branches are contracted to capillary cavities near the ends, which are plugged with cotton. The stopper and tubing must be carefully sterilized, before the test-tube is infected. This is done exactly as described above (Chamberland filter, p. 10), *i.e.*, by sterilizing the stopper in 0.1 per cent sublimate, and the tubes at 150° C., both being wrapped in paper and finally sterilized by steam after being set up in an autoclave; the stopper and tubing can, of course, be quickly sterilized together. Just before being used, the rubber stopper is unwrapped, the long tube drawn through the flame, to be sure of its sterilization, and the stopper firmly fixed in the test-tube, after which it is coated with paraffin, especially about the small tubes and where it meets the side of the test-tube. The long tube is now connected with the hydrogen generator, while the test-tube is kept at 37° C. by the use of a water-bath. According to Fraenkel, the rapid passage of hydrogen for four minutes insures the removal of free oxygen, after which the two tubes are sealed by melting at the constrictions, and the test-tube is rotated under a stream of cold water until the gelatin has solidified upon the walls. By this means a "roll-culture" of anaerobic forms is ultimately obtained (Fig. 45).

4. *The Capillary Tube Method.*—This is obviously applicable to the isolation of anaerobic species, and has been especially recommended for this purpose by Klebs and Vignal. The latter advises its combination with the passage of hydrogen and the use of solid culture-media. A small quantity of nutrient agar or gelatin is brought to the boiling point and allowed to cool off while a constant stream of hydrogen is passed through it by means of a bent Pasteur pipette inserted between the wall of the test-tube and its cotton plug (Fig. 46). By use of a water bath the gelatinized contents are prevented from solidifying. The tube is infected while hydrogen is passing through. After this has been continued a few minutes longer, and while the stream is still passing, the gelatin is quickly sucked up into capillary tubes about 50 cm. long, which are sealed at both ends and can be fastened later upon black cardboard.

C. Methods depending upon the use of the air-pump.

Where an air-pump can be had, a very simple culture-apparatus (Roux, Fig. 47) can be used. This is closed by a cotton plug and contains a rather small quantity of nutrient gelatin. Afer the latter is infected, the plug (*a*) is crowded down to the constriction *c*, the tube is softened and slightly constricted at *b*, and the gelatin remelted. By means of a piece of rubber tubing, the tube is connected with the air-pump, and the air exhausted from it, the glass being mean-

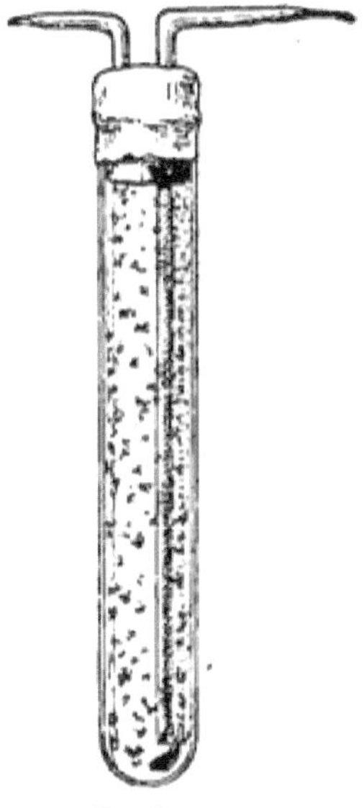

FIG. 45.

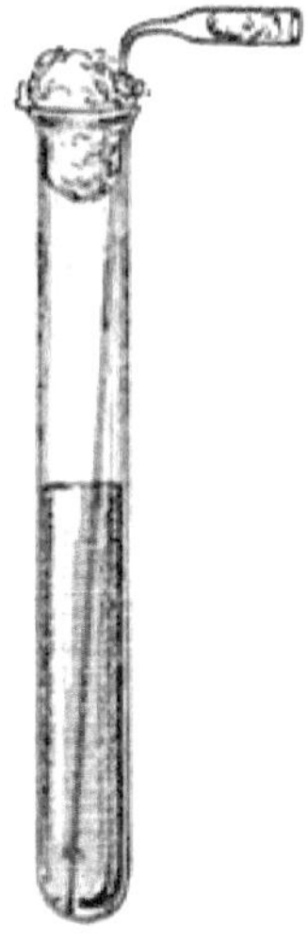

FIG. 46.

FIG. 45.—Roll-culture of Anaerobic Bacteria (Fraenkel).

FIG. 46.—Use of a Bent Pasteur Pipette for Passing Hydrogen through Gelatin Preparatory to Making Capillary Tube-cultures

time carefully warmed by passing a gas or spirit flame up and down its sides at short intervals. When the air is exhausted, the tube is melted at *b*, which hermetically seals it and at the same time separates it from the air-pump. A "roll-culture" is finally obtained by allowing the gelatin to harden in a thin layer on the inside of the tube.

According to Roux, a small water air-pump (filter-pump) is sufficient if, by means of a T-tube with two glass stop-cocks (Fig. 48), the culture-tube is likewise joined to a hydrogen generator, and the exhaustion of the air alternated five or six times with the admission of hydrogen.

Having now explained how pure material of anaerobic

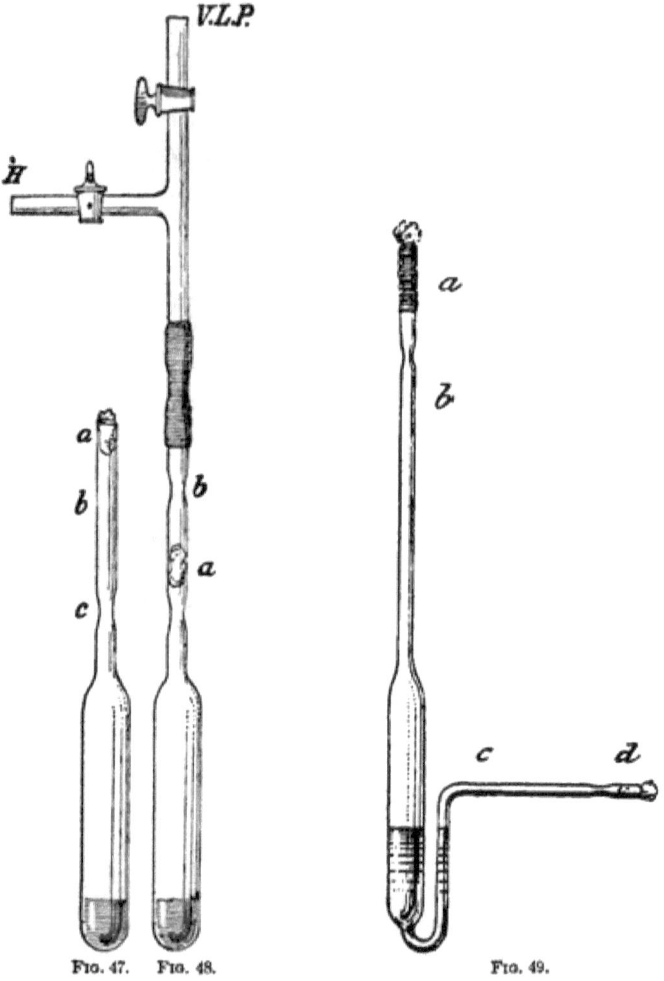

FIG. 47. FIG. 48. FIG. 49.

FIG. 47.—Culture-vessel for Isolating Anaerobic Bacteria in Gelatin in a Vacuum (Roux).

FIG. 48.—The Same Apparatus Arranged with T-tube for Alternating Connection with Filter-pump and Hydrogen-generator (Roux).

FIG. 49.—Vessel for Cultivation of Anaerobic Species under Hydrogen (Salomonsen).

forms is obtained by cultures in solid media, it remains briefly to explain the use of fluid media for the same purpose. Three courses are open, so that it is possible to employ the capillary-tube method, as described on p. 86, or the method of dilution, by adding a drop of the greatly diluted substance to each of a large number of culture-flasks, which may then be used for cultures under hydrogen or in a vacuum.

For the passage of hydrogen, culture-vessels like that of Fraenkel (Fig. 43) may be used, or my own (Fig. 49), in which the part used for the culture is about 10 cm. long and 2 cm. in diameter. The contents are inoculated through the vertical tube, closed at top by a cotton-plugged rubber tube (*a*),

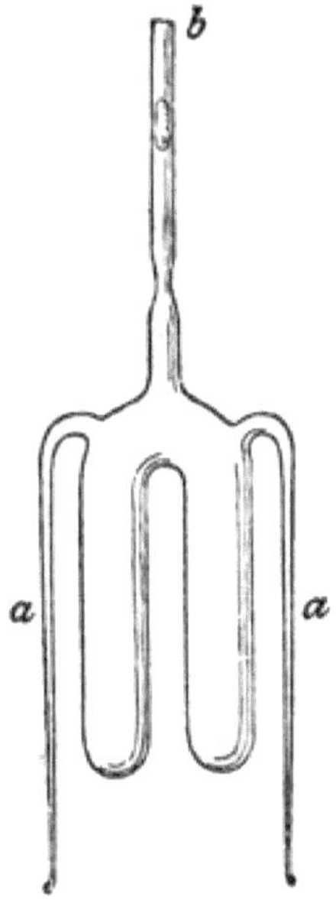

FIG. 50.

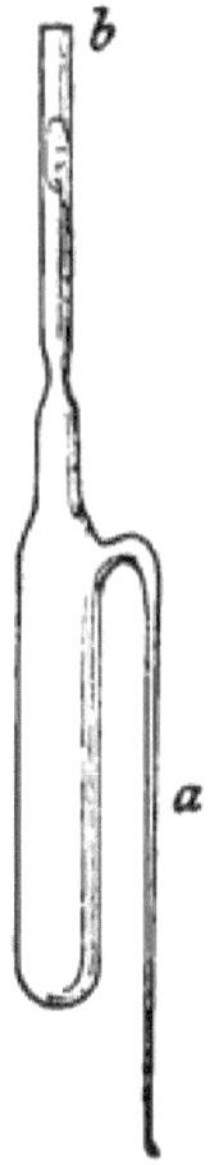

FIG. 51.

FIGS. 50 and 51.—Two Culture-vessels after Pasteur.

and hydrogen is admitted through the bent tube, provided at *d* with an ordinary cotton plug. Just behind each plug, the tubes are narrowly constricted just before the hydrogen is passed through, and as soon as the stream is discontinued the apparatus is hermetically sealed at these points. If the tubes are made long enough in the first place, these vessels may be used several times in succession.

Vacuum-cultures may be made in Roux's tubes (Fig. 47), or in those shown in Figures 50 and 51, which have been constantly used in Pasteur's laboratory for many years. These are filled through the capillary tubes (*a*), by applying suction at *b*, after which the fine tubes are hermetically sealed and the hole sterilized. The contents are inoculated by sucking some of the questionable material in through the tube *a*, after this has been heated to redness and bent away from the larger tube, and its end flamed and broken off, to be subsequently melted together again. The apparatus is connected with the air-pump by means of rubber tubing. Figure 50 shows a similar vessel in which two cultures may be carried on simultaneously, or first in one receiver and later in the other, by carefully pouring a drop into it from the first.

II. Preservation of a given Anaerobic Culture.

For keeping a pure culture upon solid media, all of the appliances recommended above may be used, and isolated-colony cultures obtained, which, as a rule, are attended by no inconvenience. If thrust-cultures are desired, other methods must be employed, such as the following four:

1. *Use of Oil or a Gelatinizing Plug.*—The inoculation-thrust is made as usual in a tube of agar or serum, but with a glass needle in preference to platinum (p. 45). Immediately afterward, a layer 5 cm. thick of olive-oil previously sterilized by boiling, or of sterilized 2-per-cent agar which has been allowed to cool to about 40° C., is poured into the tube, with the usual precautions. The advantages and disadvantages of each of these substances are self-evident.

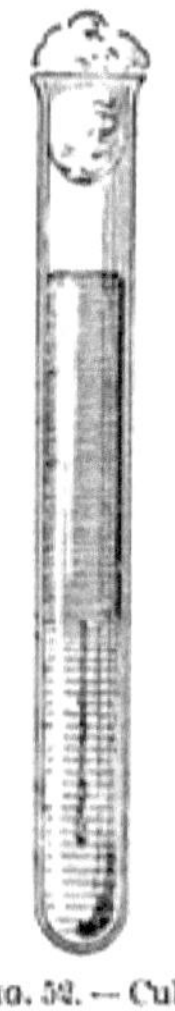

Fig. 52. — Cultivation of Anaerobic Species under Oil or Agar.

2. *Use of Entirely Filled and Hermetically Sealed Tubes* (Roux).—A tube of the form shown in Fig. 53 *a*, is plugged with cotton at its upper end, and the lower capillary portion is sealed by fusion, after which it is sterilized by hot air at 150° C. Sterile nutrient gelatin or agar is brought to the boiling point, and as soon as it has ceased boiling the opened capillary end of the tube is dipped into it, and the tube is filled up to the constriction, by suction.

By pressing a finger tightly over the upper end, and quickly raising the tube into an oblique position, the gelatin is prevented from running out of the end, which is at once sealed in the flame. The upper end is likewise sealed by fusion at the constriction just above the gelatin. When the contents are to be inoculated, one end is opened, a thrust is made with an infected fine glass needle, and the tube is again sealed. When the developed culture is to be examined, it is best to open the end opposite that from which the thrust was made, otherwise the colonies can easily be forced from the tube by the pressure of gas generated during their growth.

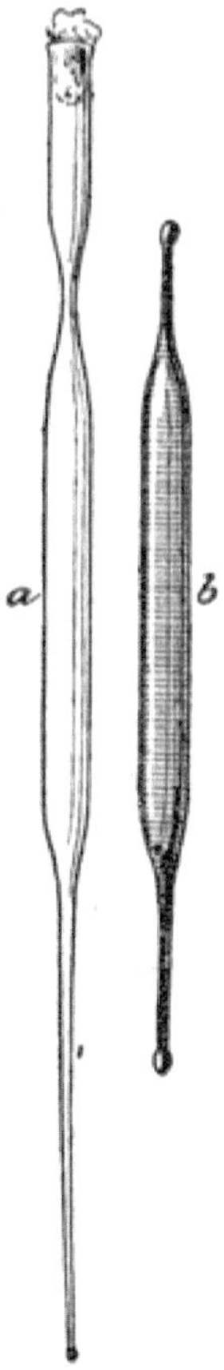

Fig. 53. — Pipette Tubes for Cultivating Anaerobic Species (Roux).

3. *Removal of Oxygen by the aid of Aerobic Bacteria.*— Roux's method is as follows: A suitable quantity of nutrient agar is boiled in a cotton-plugged test-tube, and quickly cooled in cold water. Immediately after it has become solid, it is inoculated by a glass needle, after which a small quantity of melted but not too hot nutrient jelly is poured into the tube, and, as soon as it has solidified, a couple of drops of *bacillus subtilis*, or some other decidedly aerobic species, are sown upon its surface. The test-tube is then fused together at top, and set in the brood-oven. *B. subtilis* grows rapidly upon the surface, using up the oxygen above it, and so preventing this from reaching the agar, in which the anaerobic form is consequently permitted to grow undisturbed. To obtain material for transfers from this culture, the tube is opened by breaking off its bottom, to avoid the admixture of *B. subtilis*.

Cultures can also be made in the apparatus shown in Figure 54, where the mixture of the two forms is entirely avoided. The inner tube is filled by means of a Pasteur pipette with agar-gelatin, and the space about it, with bouillon; the former is inoculated with the anaerobic form, the latter, with the aerobic species, and the outer tube is sealed hermetically at *a*.

4. *Removal of Oxygen by Pyrogallic Acid.*—Buckner has recently recommended the following method: The anaerobic form is sown in a small cotton-plugged test-tube containing agar, immediately after the latter has been boiled and rapidly cooled down. This tube is then supported on a simple wire stand (Fig. 55) [or a piece of spiral spring] in a larger test-tube, the lower part of which contains an alkaline solution of pyrogallic acid (1 gm. of pyrogallic acid; 10 cc. of 10-per-cent solution of caustic potash). The larger tube is carefully closed by a tightly fitting rubber stopper, which is slightly moist-

FIG. 54.

FIG. 55.

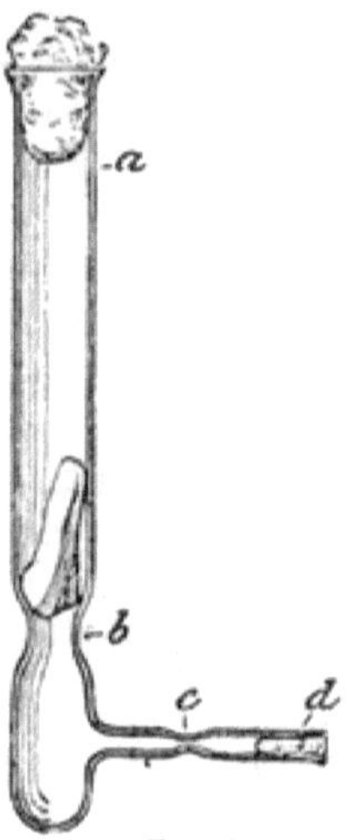

FIG. 56.

FIG. 54.—Method of Cultivating Anaerobic Bacteria by Aid of Others which Exhaust the Oxygen (Salomonsen).

FIG. 55.—Apparatus for use of Pyrogallic Acid (Buchner).

FIG. 56.—Arrangement for Cultivating Anaerobic Species on Potato (Roux).

ened on the sides, and the culture is usually set in the brood-oven at a little over 30° C. Buckner has satisfied himself that the development of so markedly anaerobic a form as the bacillus of malignant œdema is easily effected in this apparatus, though a little slowly. Absorption of the oxygen is hastened by forcibly shaking the tube now and then.

5. *Potato-Cultures in a Vacuum.*—Roux employs the apparatus shown in Figure 56 for this purpose. After the piece of potato has been sterilized in the steam-cylinder, and the condensed moisture has collected at the bottom of the tube,

below the narrowed portion *b*, the potato is inoculated in the usual way, and the test-tube hermetically sealed at *a*. The small lateral tube is then connected with the air-pump, the exhaustion is kept up for several minutes, so that all air may escape from the interior of the potato, and the tube is sealed by melting at *c*.

The preservation of anaerobic pure cultures in fluid media is exposed to no difficulties. The apparatus adapted to the passage of hydrogen (Fig. 49), or to the use of the air-pump (Figs. 47, 50, 51), is employed.

CHAPTER IX.

CULTIVATING MICRO-ORGANISMS UNDER THE MICROSCOPE.

If it is desired to follow the development of a microbe by uninterrupted observation for some hours or days, so that growth, fission, the formation and germination of its spores, etc., may be directly seen, the culture must be made in a "moist chamber," attention being given to the usual precautions, such as preliminary sterilization of the several parts of the chamber, rapid inoculation, etc. The chambers must be hermetically sealed by melting, using cement, or in some such manner, to prevent evaporation or infection. They must, therefore, be large enough to contain a sufficient quantity of air in addition to the necessary nutrient substance. Investigations of this sort are most frequently attended with some difficulty, especially for beginners, who can best practise them upon larger forms, such as yeasts and moulds, passing from these to the spore formation and germination of the large species of bacillus.

The following rules are observed in employing the various moist chambers, the most important and useful of which will be described.

Cover glasses are cleansed in the usual manner with hydrochloric acid, followed by alcohol and finally by distilled water. After careful drying they are laid in a pair of small glass trays (Fig. 13) which are wrapped in paper and, with their contents, sterilized in the dry oven at 150° C. For further certainty, the cover-glasses may be drawn several times through the flame, immediately before use. In the same way, the different kinds of slides and chambers are sterilized by dry heat, wrapped in paper, in which they remain until the moment of use.

In preparing the moist chambers it is well to use a larger

piece of sterilized paper to unwrap the slides upon, and to cover them with small bell-glasses, leaving them uncovered for only the shortest possible time. Cover-glasses must be treated with the same care. The larger shallow glass trays (Fig. 12) serve well as bell-glasses.

Several chambers are commonly employed at the same time, since some of them are liable to contamination while being prepared, even though they are sown quickly and carefully.

Cell-cultures are always started from young and strong cultures, usually in the same medium that is used in the moist chamber. Care is to be taken that the germs are uniformly distributed, and in proper number, one to four in each field of the microscope being right. As this depends upon repeated preliminary examinations, three test-tubes are used,

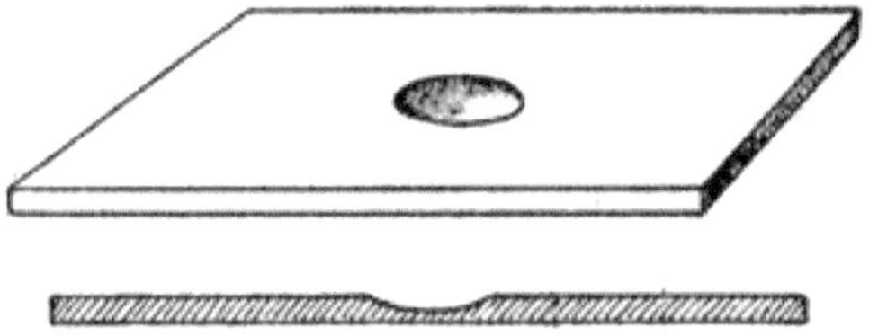

Fig. 57.—Hollow-ground Slide for use with Hanging Drops.

each containing a few grams of a suitable nutrient medium. A rather large number of germs are suspended in the first, and the effort is made by strong shaking to isolate them and distribute them uniformly. A drop is then transferred by means of a platinum loop to the second tube, and after thorough agitation the number of germs in the new mixture is determined under the microscope by examining a drop of it in the sort of moist chamber to be employed for the culture, and with the power to be used for this. If there are too many germs, a little fluid is added from the third test-tube, but if too few are present, their number is increased by adding a drop from the first tube.

One of the simplest and most frequently used moist chambers (in which Koch first observed the growth and spore formation of *Bacillus anthracis* outside of the animal organism, in 1876) is the hollow-ground slide—a glass slip of the form and size of an ordinary microscope slide, with a circular depression at the centre, about 15 mm. in diameter (Fig. 57).

After placing a small drop of culture-fluid, containing the microbe to be studied, on the lower side of a cover-glass, this is laid over the hollow and fastened to the slide in an air-tight manner by smearing with vaselin. Aside from the general rules given above, attention is to be given in the use of this cell, and in examination in hanging drops to the following: The drop must be spread out in as thin a layer as possible, because it must usually be capable of examination through its entire depth. Often it is well to place a drop of sterile nutrient fluid on the cover-glass, and to infect it at some easily recognized point along the margin of the drop.

It is best when the drop has been placed on the cover, to fix this to the slide by smearing a little vaselin around the depression with a sterile glass rod, after which the slide is lifted and turned over, and pressed against the cover-glass so that the edges of the latter are covered by the vaselin, while the culture drop projects into the hollow. Instead of fluid, a small quantity of nutrient gelatin may be placed on the lower side of the cover-glass, making a miniature plate-culture.

The preparation is first examined with a low power to get a notion of the germs present, and to find a place where they are especially well distributed, when the slide is fastened on the stage of the microscope and the germs that have been picked out are found with the higher power it is wished to use for the continuous observation. When a hanging drop is to be found and brought into focus under the microscope, this is greatly facilitated by the fact that moisture at once condenses upon the cover-glass about the drop, so as to give the cover a roughened appearance. By first finding this, and then moving the slide until it disappears, the margin of the drop is reached. Cultures of this sort can be kept on the stage of the microscope for days without suffering harm, and the development of a single germ can thus be followed, step by step.

If it is desired to hasten the process, or if the temperature of the room is too cold for the development of the bacteria, the moist chamber may be placed on some form of warm stage (such as Schultze's, Ranvier's, Israel's, or Vignal's), or the thermostat for the microscope, recommended by Panum (*Nord. Med. Arkiv*, 1874, VI., No. 7), may be used. Usually, however, it is sufficient to place the moist chamber in the brood-oven, removing it now and then for microscopic examination, care being taken to find the same part of the preparation each time. The simplest "finder" for marking a given point so that it can quickly be found again under the microscope, is that indicated by Hofmann, viz., scratching a cross on each side of the opening in the stage, one upright (+), the other oblique (×). When the point to be marked is centred in the field, two crosses are marked in ink on the slide exactly over those on the stage.

Boettcher's moist chamber (Fig. 58) consists of a deep glass ring fastened to a slide by a cement which is not injured by heating to 150° C. for sterilization. The experimenter can easily fasten the rings by one of the cements for mending glass-ware; Hansen recommends for the same purpose a solution of gelatin in glacial acetic acid, to which finely pulverized potassium bichromate is added just before use.

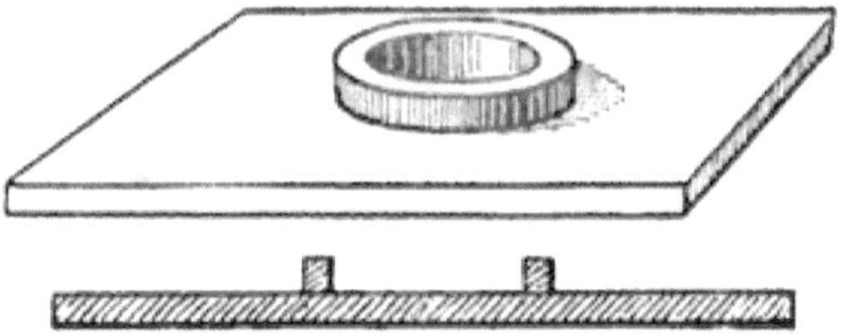

FIG. 58.—Boettcher's Moist Chamber.

A little sterilized water is placed in the bottom of the cell, and the upper edge of the ring is painted with vaselin to fasten the cover-glass. The chamber is used like the hollow-ground slide, either for examination in hanging drops, or for the observation of small plate-cultures in gelatin. The latter is the method used by E. Chr. Hansen in obtaining pure cultures of yeast for use in breweries, starting from a single cell. As a rule he employs rather large rings (about 25 mm. inside diameter), and round cover-glasses of corresponding size. A drop of nutrient gelatin containing the yeast is spread upon the latter, with all of the precautions indicated above ("Meddelelser fra Carlsberg Laboratoriet," ii., 152).

The use of hanging drops is attended with certain difficulties. The depth of the drop is relatively great, its lower surface is not plane, and the germs easily change position in it. These difficulties are partly avoided by the use of gelatin, which, nevertheless, can by no means entirely replace fluid culture media because of its influence upon the movements and mode and rapidity of growth of bacteria. Brefeld's film-cultures, described below, avoid these, but are attended with others.

A simple means of remedying some of the defects of the hanging drop consists in covering the lower surface of the drop with a fragment of cover-glass, which adheres by capillary attraction, insuring a flat bottom to the fluid. The same end is reached by the use of Ranvier's moist chamber (Fig. 59), consisting of a slide with a deep groove ground in the centre, surrounding a circular disk the surface of which is about 0.1 mm. lower than the rest of the slide. The drop of fluid is placed upon this lower central part, vaselin is painted around the groove, and the cover-glass pressed down upon it. In this way the drop is hermetically inclosed between two

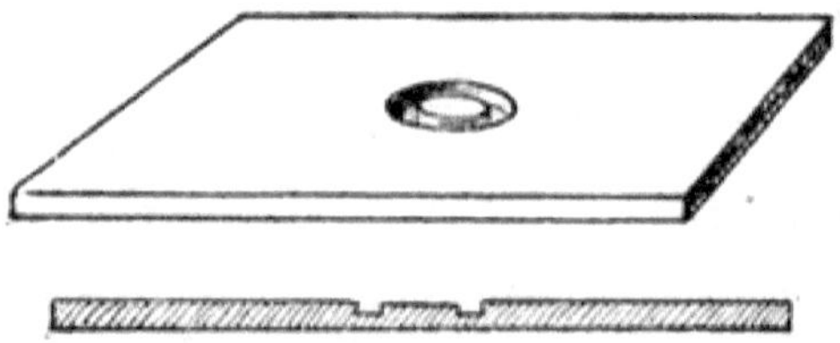

FIG. 59.—Ranvier's Moist Chamber.

parallel flat surfaces, while the groove serves as an air-chamber.

A very expensive and fragile, but otherwise excellent moist chamber, is the modification of De Bary's and Geissler's model used by Brefeld in his study of *Bacillus subtilis*, in which he succeeded in following the entire development from spore to vegetative cells and from these to new spores. This is a glass tube 20 cm. long (Fig. 60), in the middle of which a transversely cylindrical cell 2 mm. deep is blown, the flat top and bottom of which have the thickness of a cover-glass. To prepare it for use, it is thoroughly cleansed with hydrochloric acid, distilled water, alcohol, and ether, and each end of the tube is plugged with cotton. When dry it is sterilized at 150° C. The microbes to be studied are uniformly distributed in a suitable culture fluid, until by examining several drops spread very thinly, it is found that they are present in the right number (about one to four to each microscope field). One plug is now removed, the open end is quickly flamed and immersed in the fluid, which is drawn into the cell by suction applied to the other plugged end of the tube. When the chamber is quite full, the fluid is allowed to run out again, the wet end of the tube is dried with sterilized filter-paper, and replugged, both ends are sealed with sealing wax, and the cell is ready for examination. The inside of the thin-walled chamber is here lined by a very thin film of fluid containing the germs, so thin that the bacteria do not move about in it. A suitable part of the preparation is found, and the chamber is fixed upon the stage of the microscope. In searching through it, which is done with the lowest power sufficient for recognizing the germs, care is taken not to crowd the objective against the thin and fragile wall of the cells, a mishap which is rendered all the easier by the fact that the latter is not perfectly flat. It is most easily fastened on the stage by first placing it upon a thin and well-cleansed slide, to which the cylindrical ends can be attached by paraffin. Admitting of the safe and easy transmission of gases, this moist chamber can also be used for studying the development of bacteria in different gases.

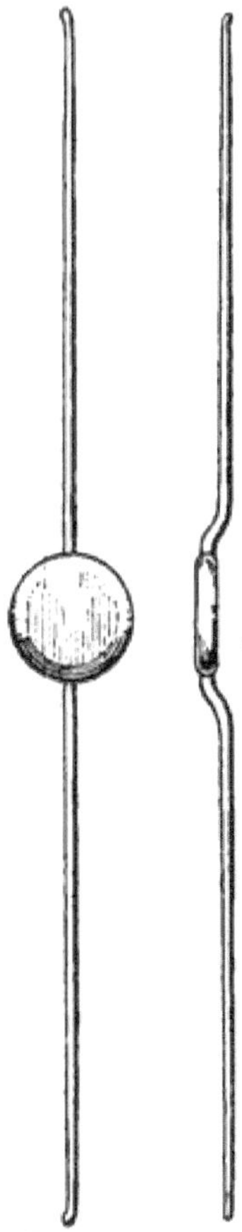

FIG. 60.—Geissler-chamber for Cultivation in a Film (Brefeld).

Other forms of moist chambers are easily improvised out of the material and appliances to be found in every laboratory. As it may be important at times to work with a very large number of cell-cultures at the same time, some of these easily extemporized forms deserve description.

Buchner studied the germination of spores of *Bacillus anthracis* in a moist chamber such as is shown in Figure 61. A very small quantity of a pure culture containing only the spores of this species is dried upon a cover-glass, and a small drop of culture-fluid added. The cover is then supported on a slide at two edges, by fragments of cover-glass, the entire margin sealed with some cement, and the slide placed on a warm stage.

A cell, the inventor of which is not known to me, consists of rather thick pasteboard (No. 16), cut into the form shown in Figure 62, and carefully sterilized by boiling or in the steam cylinder, which is best effected by piling several pieces up between a couple of microscope slides and tying them firmly together. One of the pieces of pasteboard, while still wet from sterilization, is laid on a sterile slide and pressed into contact by a glass rod. A cover-glass, with the hanging drop, is laid over the hole in the pasteboard and carefully pressed down on it, after which sterile water is added at one edge until the paper refuses to absorb more. In this simple chamber, the paper serves both as a source of moisture and a means of attaching the cover, so that it must be kept from drying out as long as the chamber is in use; consequently, when not undergoing examination, cells of this kind, like plate-cultures, need

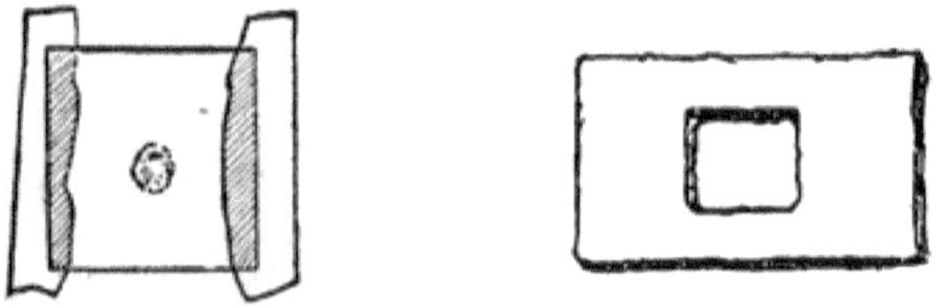

FIGS. 61 and 62.—Improvised Moist Chambers. 61, Buchner's form; 62, Cardboard for Cell.

to be kept in a large moist chamber (*cf.* p. 60), and from time to time the pasteboard must be resaturated by the addition to the outer edge of a little sterile water or very dilute sublimate solution. For this reason this form of cell is not well adapted to continuous observation under the microscope for a period of days, though it may often prove very useful as an accessory. Other materials may also be used for the preparation of moist chambers of the same sort.

In his studies of the life-history of the higher fungi, Brefeld has cultivated them under control of the microscope by the simple use of slides upon which a nutrient fluid was spread in a thin layer. These plate-cultures in fluid are not suited to the study of bacteria. So far as yeasts are concerned, they offer no especial advantage over cell-cultures, but are attended by certain disadvantages. On the other hand, they are of the greatest importance for the study of the development of higher fungi, and for this reason merit brief consideration here.

The solutions described on pages 19 and 20 are used, generally after being boiled down so that they are not too fluid. By means of a glass rod, an elongated drop is placed on a sterile slide, and enough of the spores to be studied are added and distributed with a platinum needle so that only one or two occur in each field of the microscope. It is evident that such cultures cannot be used for prolonged uninterrupted observation. If kept long upon the stage of the microscope, they will be ruined by drying out or contamination. Between the different examinations, they must be kept in larger moist chambers, the air of which is always saturated with moisture (a common flat porcelain platter filled with water and covered by a bell-glass dipping into the water), and they can scarcely be used successfully except in laboratories where the air is kept as free as possible, by great care, from germs, and especially moulds; but Brefeld's works show sufficiently how far it is possible to utilize this method. Besides other advantages over similar slide-cultures on solid media, these admit of transplanting the spores after germination has begun, more favorable conditions are given the moulds for growth and development, and it is possible to renew the supply of food material during weeks or even months, as it becomes exhausted, by adding a fresh drop of culture-fluid at the margin every day or two.

CHAPTER X.

INOCULATION OF ANIMALS.

We come now to a group of experiments to which, in a certain sense, all of the preceding chapters point—experiments calculated to give the proof that a large number of the most familiar contagious diseases are induced by bacteria, and to render possible a more intimate study of the mutual relations between these bacteria and the animal organism. In this connection it must be said that conclusive proof that a given infectious disease is due to a specific bacterian form, is given only when such a form, well characterized morphologically, chemically, or physiologically, can always be demonstrated in the organs by the microscope and by cultures, in this disease and only in it; and when, further, these bacteria, after being cultivated pure for several generations outside the animal organism, when reintroduced into an animal of the species they were originally obtained from, produce the same disease in this, and are then demonstrable in its tissues under the microscope and by pure cultures.

Various kinds of animals have been used for experiment, but usually the smaller rodents (rabbits, Guinea-pigs, rats, and mice), or birds (fowls and pigeons) are employed. It is now well known that inoculations, not of one but of a large series of infectious diseases, may be effected upon one of the most accessible, cheapest, smallest, and most prolific of mammals—the white mouse, which is more or less receptive for splenic fever, chicken-cholera, malignant œdema, several forms of septicæmia, some pyæmic affections, glanders, tetanus, etc., etc. While the white mouse (an albino of the house mouse, *Mus musculus*) is very slightly receptive for glanders, according to Loeffler, the European field mouse (*Arvicola arvolis*) is very susceptible to it (Loeffler), as is also the wood mouse (*Mus sylvaticus*) according to Kitt. On the other hand, the field mouse is not receptive for mouse septicæmia, to which the white mouse shows a great susceptibility (Koch). Both field and wood mice are easily kept in captivity if fed upon oats and moistened bread. The wood mouse is very active, and its bite quite painful; and it is not advisable to keep several individuals of either of these mice in the same jar after inoculation, because in case one dies the others devour it at once. Where it is only desired to perform the most important fundamental experiments in bacterial infection, it is, therefore, usually sufficient to employ white mice.

Keeping Mice.—After many trials, I can highly recommend the use of the cracker-boxes spoken of on p. 2, as cages for mice. A large number of holes about as large as a half dime are made in the cover, *e.g.*, nine rows, of nine each. In case the holes are not large or numerous enough, evaporation from the interior of the box is checked too much, and the mice cannot stand the resulting humidity. The box is filled nearly half full of sawdust, a little cotton is laid on this, and the cage is ready for use. The mice are fed upon white bread softened with water and oats. The sawdust contributes largely to lessening offensive odors, and, in connection with the cotton, grain-hulls, and the excrement of the animals, forms a warm, soft, and dry mass, in which the mice tunnel, and thrive and breed well. Ten or fifteen adult mice can easily be kept in such a box, or a still larger number of young ones. Emptying the box and renewing the sawdust and cotton is only necessary once a month or even once in two or three months, depending upon the number of animals kept in it.

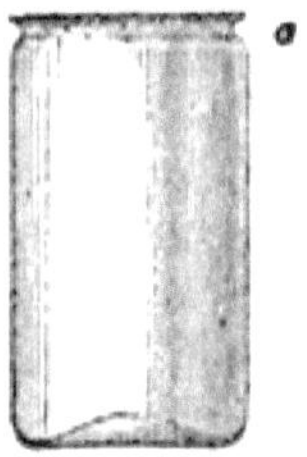

Fig. 63.—Mouse-jar.

For isolating infected mice, it is best to use common glass pickle-jars, holding two quarts (Fig. 63). These jars are readily disinfected after use, and the animals can be easily observed in them. Each is filled for a third of its depth with sawdust, and covered at top by a square piece of close iron gauze, large and flexible enough so that it can easily be slipped over the mouth of the jar and pinched into the groove *a*. For greater security a heavy object may be laid upon the lid. In this way it is fastened securely enough so that an animal cannot lift it and escape, while it is readily taken off and replaced, and can be disinfected in the flame without difficulty.

When white mice are constantly used for inoculation, they are usually raised in the laboratory. Starting their propagation is not always easy, since mice are quarrelsome, and males which have not been raised together often fight and injure one another, while they are also apt to kill and eat the young. As a general thing, a few pairs are taken for a beginning, and kept apart. When a female is near the term of pregnancy, she is removed from the male and kept in one of the mouse-jars just described, where she remains with her young ones until their eyes are open, when they are all removed to one of the large cages, in which two females and their litters can be safely kept, as a rule, provided the young are of about the same age. The young mice which grow up together in this way generally get along so well that it is unnecessary to remove the pregnant females as long as there is room for their burrows and young.

The mode of inoculating an animal, and the point at which this is done, depend upon the object in view. If it is merely desired to infect the animal, introducing the contagium through a wound in the skin suffices in many cases. Not infrequently, however, it is wished to imitate the natural infection through the uninjured mucous membrane of the respiratory or digestive system. In other cases it may be necessary to introduce the virus under the dura mater, by trephining. In short, the most different organs and parts of organs are used for the introduction of cultures of bacteria, or infectious material. Only the most important of the methods of inoculation are considered in the following pages.

In most cases the inoculation of an animal is practically painless, so that it is anæsthetized only when some larger operation is involved, as in trephining for inoculation with rabies. Mice and rats are placed under a bell-glass with a wad of cotton moistened with ether, until, after a short period of excitement, they succumb to the anæsthetic. After they are removed from the bell-glass, and tied, insensibility is easily kept up, as they are very tolerant of ether. It is otherwise with Guinea-pigs and rabbits, the latter, especially, dying easily when anæsthetized with ether, or, particularly, chloroform. Practice, however, enables the operator to lessen the number of deaths from this cause, and in the Pasteur Institute only one to two per cent of the rabbits die from the anæsthetic, which is administered by pouring a teaspoonful of chloroform over a piece of absorbent paper folded as if for a filter, which is then held like a cornucopia over the nose of the animal. After a few seconds the respiratory movements stop, and when they are resumed shortly afterward, insensibility as a rule is complete, as soon as this occurs the administration of chloroform is discontinued.

Cutaneous inoculation is best performed on mice by abrading the ear with a vaccine point dipped in the virus, although it may be effected anywhere on the body after removing the hair. Scarification of the skin followed by rubbing the virus in, or simply rubbing the latter into the uninjured skin, may also be practised.

Subcutaneous inoculation of mice is best performed under the skin of the back, just above the root of the tail. The mouse being in a glass jar (Fig. 63), its tail is seized between the thumb and forefinger of the left hand and drawn out over the rim of the glass, while the body hangs into the jar, which is covered with a small piece of board, held in place by the free fingers of the left hand, so that there is just room for the passage of the root of the tail between it and the edge of the jar. [When the usual cylindrical jars of the German laboratories, with covers similar to those of candy-jars, but the top replaced by coarse iron netting and weighted, are used, these covers need only be tilted a little at one side to serve the same purpose. If gray house-mice are used, they are apt to be far more active than white mice, and it is best to quiet them by holding a handkerchief, moistened with a few drops of ether, over the cover, without carrying the etherization to the surgical point. In any such operations, the long "mouse-tongs" of the German dealers are a great convenience for first seizing the tail of an animal.—W. T.] The hair just above the tail is clipped off, and by means of a lancet-needle or a pair of scissors a small opening is made in the skin and a pocket torn in the subcutaneous connective tissue, into which the infectious material is placed. Solid bodies are introduced by means of the forceps; very small quantities of cultures, etc., by the lancet-needle or platinum needle; and slightly greater quantities of fluid, by means of capillary tubes or Pasteur pipettes.

When a considerable amount of fluid is to be injected into the subcutaneous tissue, a sterilized Pravaz syringe must be used. The common Pravaz syringe is not adapted to sterilization at high temperature, since the piston is packed with leather, and the tip cemented on. Generally, a syringe is used the tip of which screws on to a thread on the glass cylinder, and with an elder-pith plunger, so that it may be sterilized by either moist or dry heat (Straus and Collin).

Material from cultures in fluid or from vigorous cultures upon solid media, can easily be taken from test-tubes by the pipette or platinum loop. But when the bacteria to be used for inoculation grow in a narrow line along the needle-thrust in a test-tube, and it is desirable to have a relatively large quantity of material, one should rather cut out the entire colony and use it either with or without melting. The process is a little different according as the culture is in gelatin, agar, or serum.

A gelatin culture is dipped in warm water or warmed over the flame until the outer part melts and the gelatin cylinder becomes free from the test-tube, so that it can be shaken down against the cotton plug, when the latter is carefully removed and the gelatin allowed to fall into a sterilized watch-glass, where, with a sterile and slightly warmed knife, all of the peripheral part containing no bacteria is cut off and removed, leaving only a small prismatic piece of gelatin containing the colonies.

Agar-agar, which neither adheres to the glass so firmly as gelatin, nor melts at so low a temperature, is removed in a somewhat different manner. The test-tube is warmed somewhat, a Pasteur pipette is passed to the bottom of the tube, between it and the agar, and by a strong puff through the pipette, the entire culture can usually be blown out of the glass. If this does not succeed, a warmed glass rod, such as the handle of a platinum needle, is passed down to the bottom through the agar at one side, so as not to come in contact with the colonies of bacteria. As soon as it has cooled, it can be used to draw the agar out by a series of pushes and pulls. The agar being removed into a sterilized watch-glass, it is treated just like gelatin.

Serum is far harder to manipulate, because it adheres to the glass and cannot be removed by melting, so that it is necessary to cut the colonies out of the test-tube as well as possible with a long-handled pointed knife.

Intravenous injection cannot be resorted to with mice, because of their small size. In the case of larger animals, it can generally be effected without exposing the vein by simply thrusting the needle of the syringe through the skin into the cavity of the vessel.

With rabbits, it is particularly easy to inject material into the veins of the ear and leg. For the former, the rabbit is wrapped tightly and carefully in a long towel, so that only the head projects. The ear is washed with 2-per-cent carbolic acid, partly to disinfect it, and partly because the vessels are more easily seen when the hair is wet. An assistant holds the animal's head and compresses the base of the ear, so that the

FIG. 64.—Injection of Culture into the Vein of a Rabbit's Ear.

veins swell, the ear is seized between the thumb and forefinger of the left hand so that it is slightly tense over the side of the forefinger, and the canula is carefully thrust through the skin into the vein (Fig. 64). The assistant now releases the end of the vein, the operator holds the canula still in the vein with his forefinger and thumb, and injects the fluid slowly. Bleeding is easily stopped by brief compression, or by amadou. In case the vein has not been pierced, the fluid soon begins to distend the surrounding connective tissue, when it is almost always useless to withdraw the canula a little way and make another effort to force it into the vein, but another vein is usually at once chosen for injection. Commonly the branch of the vein running along the posterior edge of the ear is selected, but there are many other available points on the ear.

When the ear is used for injection, there is the further advantage that one may be sure of really intravenous inoculation without attendant accidental inoculation of the wound, by simply removing the portion through which the injection was made, by a quick clip.

By connecting the needle-shaped canula with a larger syringe by a rubber tube, large quantities of fluid may be introduced into the veins of the animal, in the same simple and painless manner. As the injection must be effected very slowly, and in this case requires a longer time, it is best to fasten the animal upon a "rabbit-board" (the French model, used in Marey's laboratory, is far preferable to Czermak's) so that a sudden movement of its head shall not displace the canula.

For injection into the leg veins, the rabbit is firmly and carefully wrapped in a long towel so that only one hind leg is left free, the animal being allowed to draw the other up under its body. An assistant sits with the rabbit upon his lap, one hand extending the free leg firmly, while with the other he grasps and compresses the thigh a little above the knee-joint, so as to cause the veins to swell, his fore-arm resting upon the body of the animal. One of the subcutaneous veins in the lower joint of the leg is found, the hair over it is clipped off, the skin washed with 2-per-cent carbolic acid, and while the leg is further steadied by the left hand, the canula is pushed through the skin into the vein, where it is held by the pressure of the left thumb.

Intraperitoneal inoculation is most conveniently effected with a sterilized Pravaz syringe. The entire thickness of the abdominal wall is pinched up in a longitudinal fold, through which the canula is forced crosswise so as to emerge on the other side. On releasing the fold, the canula is carefully withdrawn enough, so that its point lies within the body cavity, while there is no danger of injuring the intestines.

In case large quantities of fluid are to be injected into the peritoneal cavity, a graduated glass pipette may be used, one end of which is plugged with cotton, while the other is connected by rubber tubing with a canula similar to that of the Pravaz syringe. The canula is inserted in the manner indicated, and the fluid blown in. The same kind of inoculation can also be effected with a Pasteur pipette, the perforation being made with a lancet-needle, beside which, as a guide, the pipette is passed into the abdominal cavity. Mice and rats are best etherized before inoculation.

Inoculation in the peritoneal cavity is especially important in admitting of the introduction of very considerable quantities of either fluid or solid substances into the animal. In the latter case, the body cavity must be opened for a greater distance along the linea alba. By such laparotomy, carefully performed, an entire organ from a larger diseased animal may easily be inserted into a smaller one, *e.g.*, the heart and kidneys of a rabbit, into a rat.

Inoculation into the anterior chamber of the eye acquired especial importance in the study of tuberculosis, since by inserting small masses of tuberculous material into the eye Conheim and Salomonsen succeeded in inducing tubercle of the iris, by which it became possible to directly observe the incubation period of miliary tuberculosis, and its independence of a preceding suppurative process. It has also recently been used for rabies inoculations.

The operation is performed by fastening the rabbit, on its belly, usually on an operating board, where its head can be perfectly fixed. An assistant opens the eyelids, most conveniently by sitting so as to face the operator, grasping the animal's head with both hands. Occasionally the nictitating membrane is so large that it must be held back. With a pair of forceps, a fold of the conjunctiva is seized, and with a suitable knife a cut 2 to 3 mm. long is made in the cornea, near its margin, with the usual precautions observed in ophthalmological operations. If the material to be introduced is a solid substance, it is passed through the opening by means of slender curved forceps, usually with very fine smooth points opening parallel, and by carefully stroking the cornea with a Daviel's spoon, the effort is made to crowd the introduced material into the bottom of the anterior chamber. If this is not easily effected, it is allowed to remain where it lies. Fluids are injected through a (usually) blunt and bent canula, inserted into the cut. To anæsthetize the eye for the operation, a 2-per-cent cocaine solution is dropped into it, the maximum action being reached after fifteen minutes (Howe).

Another place, easily accessible for observation, which has often been used for inoculations, is the cornea, in which the conditions are very favorable for observing the effect of bacteria upon the connective-tissue cells. By means of a blunt needle, a large number of pricks or scratches are made in the corneal tissue, without perforating it, and the bacteria are gently rubbed into the wounds; or the needle may first be dipped into the virulent material,—but the first method seems to give surer results.

Inoculation beneath the arachnoid, as first used in hydrophobia infections in Pasteur's laboratory (Pasteur and Roux), is effected on rabbits and guinea-pigs in the following manner: A cut is made through skin and aponeurosis, on the crest between the eyes and ears, and the edges of the wound are held apart by a small tenaculum. The trephine (about 5 to 6 mm. in diameter) is applied behind the orbit, on one side of the middle line. When the circular groove in the bone is deep enough, the centre pin is withdrawn so that it shall not injure the dura mater. As a general thing, the fact that the bone is cut through, is observed in time, but for greater certainty it may be seen from time to time whether the disk of bone cannot be lifted out. When the dura mater is exposed, it is pierced with a needle-shaped Pravaz canula, bent nearly at right angles, which is drawn slightly toward the operator to avoid injury to the underlying brain, and two or three drops of the diseased spinal cord infusion are injected. The wound is cleansed with 2-per-cent carbolic acid and closed by a couple of sutures. In the case of dogs, the skin is pushed aside after the cut is made, the temporal muscle is loosened, and the skull is trepanned in the fossa temporalis, where the skull is thinner, and there is little bleeding.

As compared with all other methods employed in the study of rabies, inoculation under the dura mater has the advantage of giving absolutely certain infection, as well as the shortest and constant incubation. But according to the latest investigations of Roux and others, the far simpler inoculation in the anterior chamber of the eye gives as certain infection as the subdural.

Pure virus is always found in the medulla, etc., of animals dead of hydrophobia, which is best suspended in sterile bouillon or 0.7 per cent salt solution, and injected with a Pravaz syringe. The removal and preparation of the material must naturally be effected with the greatest possible cleanliness, to avoid accidental infection. In the Pasteur Institute the material for inoculation is prepared by snipping out a small piece of brain substance from the bottom of the fourth ventricle and placing it in a wine-glass holding about 150 cm., which has been covered with paper and sterilized at 150° C. Here it is rubbed to a soft paste by means of a glass rod, and a small quantity of sterile bouillon is gradually added, while it is constantly stirred. The glass is then covered with paper and set aside until the coarser particles have fallen to the bottom, the upper, slightly turbid, layer of fluid being used for injection.

The medulla oblongata retains its virulence unchanged for at least a month when kept in pure neutral glycerin—a fact which should be remembered whenever the occasion arises for sending parts of the central nervous system of man or one of the lower animals to a distant laboratory for diagnosis.

Rabies vaccines are prepared as follows (Pasteur and Roux): A rabbit weighing two kilos and measuring about 45 to 50 cm. from the nose to the root of the tail, is inoculated by trepanning with "virus fixe." Six or seven days later it shows symptoms of rabies, and dies on the tenth day. Before putrefaction sets in, the central nerve system is removed as follows: The skin is split down the back from nose to tail, dissected to one side, the dorsal and cervical muscles are loosened from the spinal column and cranium, and the spinous processes are removed by bent shears. The nose is then seized with a pair of strong bone nippers, and so held fast with the left hand, while the theca cranii is snipped and broken off by means of a pair of Liston's bone scissors. Then the vertebral arches are removed one by one, by snipping them through on the right and left side, as near the body as possible, without injuring the spinal cord, and breaking them off. This is more easily described than done, for the spinal cord is easily crushed, especially in the cervical region; but with practice one gradually learns to expose it for its entire length without injuring it with the coarse instrument used. Then the cord is cut off over the cauda equina, its membrane is seized with forceps just above the cut and it is raised for a distance of 6 to 7 cm., and all adhesions are cut. The loosened piece is cut off, a sterile silk thread tied about one end, and hung up (Fig. 65) in a litre flask with two openings previously plugged with cotton and sterilized at 150° C. Enough caustic soda to cover the bottom is dropped into the flask, the spinal cord is hung from its neck, and the whole set aside at 20 to 25° C. The vaccines are prepared in the manner described above, from such pieces of spinal marrow, hung up and dried for a longer or shorter time.

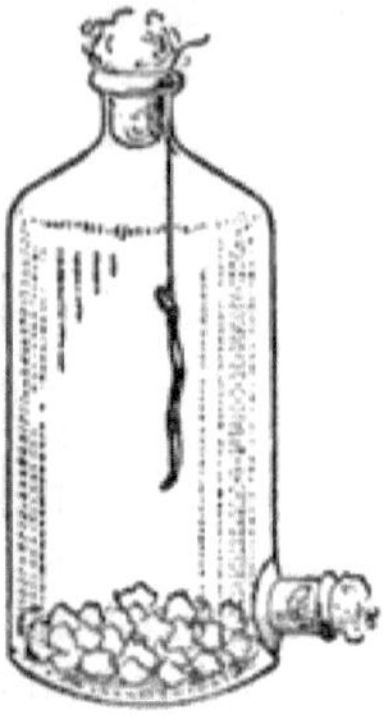

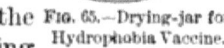

Fig. 65.—Drying-jar for Hydrophobia Vaccine.

Infection through the digestive tract is effected by carefully mingling the contagium with some food that the animal is fond of, which is given it in not too large quantity, the animal being previously kept without food for a day or less, if necessary. In case of mice and other small animals, this is the only available method. In the case of larger mammals (rabbits and Guinea pigs), simple feeding is also used when large quantities of solid material are to be introduced into the stomach, and it is the best means of inducing infection through the alimentary canal, because it is in this way that natural infection by means of food occurs as a rule. Still it cannot always be employed, and other means must then be resorted to. Fluids (*e.g.*, cultures of bacteria) can likewise be poured or injected through the œsophagus, the rabbit being wrapped in a towel with the exception of its head, and held on the lap of an assistant. The animal is made to open its mouth by slight pressure upon its cheeks over the molar teeth, and a small wooden gag (Fig. 66) is inserted so that its incisors rest in the grooves on the upper and lower surfaces (*b*). Through the perforation in the gag, a No. 17 catheter can easily be introduced into the rabbit's ventricle. Adhering bits of food show that it has been inserted right. In case of the Guinea pig, a smaller catheter (usually a bulb catheter) must be employed, and it requires to be passed down with especial caution.

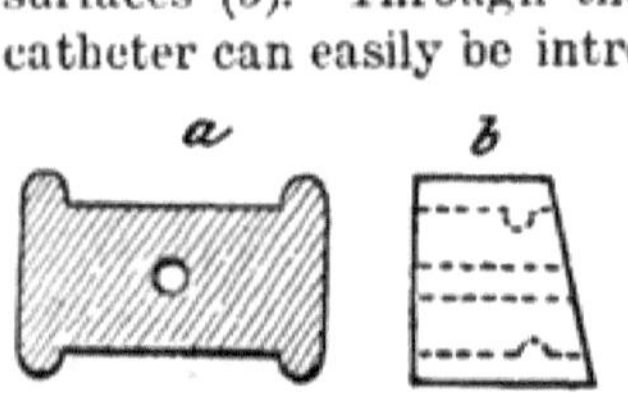

Fig. 66.—Wooden Gag for Passing a Catheter into the Stomach, in Feeding Experiments. *b*, Section of same.

Solids are quickly introduced into the crop of doves or fowls by opening the beak, and placing the food in the mouth, as far back on the tongue as possible, when the bird quickly swallows it. Small meal pellets are formed with cultures and other fluids containing bacteria, and these are introduced in the same way.

The introduction of infectious fluids into the stomach and intestine through the abdominal wall, by means of a Pravaz syringe, is more certain and reliable. If the contagium is to be introduced directly into the intestine, the abdomen must be antiseptically opened, a fold of the intestine found and carefully fixed, and the canula thrust through its wall. In this way cultures of the cholera spirillum are injected into the duodenum (Nicati and Rietsch), so that they do not pass through the stomach, and the action of the gastric juice upon the microbes is avoided.

Pathogenic bacteria can be brought in contact with the intact surface of the lungs by injection through an opening made in the trachea. For the sake of entire certainty that the virus is not at the same time introduced through the open wound, this may first be allowed to heal, leaving a small tracheal fistula. In the case of large animals, Arloing, Cornevin, and Thomas have reached the same end by passing a drainage tube through a metal canula, and injecting the desired material through the former, in small quantities, so as to avoid spasms of coughing.

If the natural means of infection through the mucous membrane of the respiratory organs are to be imitated as closely as possible, the bacteria must be very finely and uniformly suspended in the air inhaled by the animal, so as not to cause mechanical irritation. This is best effected by use of a spray or dry powder. The former method is open to the objection that the spray contains not only extremely fine drops, but likewise some which are a little larger, so that the mucous membrane of the nose, or the entire animal, becomes quite damp. To avoid this, Buckner joins the atomizer to a two-necked Woulf flask holding about three litres, where the coarser drops are allowed to settle, while the finer, as a nearly imperceptible mist, are passed out of one neck of the flask through a tube to the inclosed chamber where the animals are.

For the second method, a fluid-culture of the bacteria is poured over a rather large quantity of spores of *Lycoperdon giganteum*, which are then dried off over calcium chloride, and the powder is blown into the closed animal cage by the bellows (Buckner). In his experiments on the entrance of spores of *B. anthracis* through the lungs, Buckner used 0.25 gm. of powder for a space of three litres, and allowed the inhalation to continue ten to fifteen minutes. Other powders (carbon, talc, etc.) can also be used as a vehicle for cultures of bacteria, but the puff-ball spores are superior to all of these in being very small, uniform, and of low specific gravity. In case the bacteria are killed or deprived of their virulence by drying, this method, of course, cannot be employed.

In such experiments, the animals must be untrammeled so that their respiratory movements are not checked or altered, and it is best, therefore, to let them remain unfastened. A number of mice can be placed together loose in a tight tin box with a few breathing holes plugged with cotton. If it is necessary to fasten the animals so as to keep their heads in a given direction, wire gauze can be used, as it may be bent and shaped at will. A single mouse can conveniently be placed in a glass tube 3 to 4 cm. in diameter, with a bored and split cork in one end, which serves to hold its tail.

When inhalation experiments are performed with bacteria that are pathogenic for man, the operator should remember to take sufficient precautions, such as having tight vessels and tubes, cotton-plugged air-spaces about the cages of the animals, and long tubes for the atomizer, and the work should be done in the open air.

CHAPTER XI.

CULTURES FROM MAN AND ANIMALS. COLLECTING AND PREPARING PRIMARILY STERILE CULTURE-MEDIA.

In attempting to obtain cultures of the pathogenic micro-organism from the blood or organs of an animal which has died from an infectious disease, the rules given in Chapters I. to III., and V., are to be observed, as well as the following:

Since the more or less dirty hair or feathers of the animal contain a possible source of contamination, it is best to skin or pluck the body before opening it. All of the parts covered by hair may also be washed with sublimate solution, or covered with filter-paper wet with sublimate.

All instruments used for preparing the body must be carefully sterilized. Since repeatedly heating them to a high degree in the flame works destructively upon knives, scissors, and forceps, it is best to have two sets of these instruments; one intended for the coarser part of the work, sterilized by flaming, and commonly used so hot as to cause the flesh to hiss while the cut is being made; the other for finer operations, first sterilized at 150° C., in a metal case or wrapped in paper, and merely hastily flamed just before use.

In addition to the gas or spirit flame, a dish of 0.1 per cent sublimate solution, and of 5-per-cent carbolic acid, must be kept ready for the disinfection of hands and instruments. To avoid contamination from contact, the sterilized instruments are laid upon glass supports (Fig. 18) and covered with a bell-glass, or dropped into a short cylindrical vessel with their points projecting out at top.

While exposing the organs from which material is to be taken for cultures, use is made of strongly heated instruments, which, as far as possible, are kept from touching the organs. No definite rules can be given, further than this:

If a culture is to be started from fluid contained in a cavity, it is best to singe the surface over, before piercing the wall. For example, if a collection of fluid in the pleural cavity is to be used, after exposing the ribs and intercostal muscles in the usual way, a part of one of the intercostal spaces is scorched by means of a glass rod heated in the flame, and then, with a glowed needle or strongly heated small knife, the pleural cavity is opened, and the small gaping wound made for inserting the inoculation-needle, capillary tube, or pipette, according to the quantity of fluid needed. In the same manner, blood is obtained from one of the anterior chambers of the heart. After cauterizing the surface, as a rule the capillary tube or pipette may be thrust through the thin wall, without the previous use of a knife, which only becomes necessary when a thin and flexible platinum needle is to be inserted.

To obtain material from the substance of a solid organ, one of the thicker and more rigid needles is used. The organ may be snipped in two with scissors carefully flamed or even glowed, and the needle thrust into its substance from the uncontaminated cut surface, or it may be torn open with sterile fingers or forceps, and in this way a surface exposed which has not been touched at all by the instruments. [Loeffler advises that when parts of cut organs are to be broken in this way, the surface be first disinfected by dropping the whole piece into 0.1 per cent sublimate, and moving it about in this for a minute or two]. By using flamed, but not too strongly heated, curved scissors, small pieces of the organ may be snipped out and placed on or in the culture medium, at least after being crushed between two sterilized glass slides. This method is recommended by Koch in sowing from tubercle.

Since the white mouse is the animal presumably used in the elementary inoculation experiments, the following description is given of the best means of obtaining cultures from its heart blood. As soon as the external examination is complete—especially about the inoculation wound—the body is extended upon its back by a shawl pin through each leg, on a small board, which it is well to first cover with a sheet of paraffined paper over which a couple of sheets of filter-paper are laid to absorb blood, etc. The head is finally fixed by means of a fifth pin through the nose. The skin is dissected away from the ventral surface as far as possible, and entirely cut off, the occasion being taken to examine the axillary and inguinal lymphatics, and, if necessary, to obtain blood from the axillary vessels for microscopic investigation. Seizing the prominent ensiform process with a pair of sterilized forceps, it is lifted strongly, and by means of flamed sharp-pointed scissors, one end of which is inserted under the ribs, a sufficient part of the thoracic wall is removed to expose the pericardium. In this way the intestines are not approached, and the instruments are not exposed to infection from this dangerous source. Care is taken to avoid touching the anterior surface of the heart with the scissors. With two pairs of small flamed forceps the pericardium is now torn open, and a small cut is made into the exposed heart. Usually sufficient blood to wet the platinum needle at once bubbles out of the wound; otherwise the needle is pushed through the opening, and the culture started from the fluid obtained. The dissection is now completed in the usual way, but with especial care and cleanliness so as to avoid rendering it difficult or impossible to make cultures from other organs in which morbid changes may subsequently be discovered. When the dissection is finished, and the necessary preparations have been saved, the cadaver is wrapped in the filter-paper it lies on, and burned up.

The manner of making the dissection naturally varies with the organ chosen for starting a culture, the rule being to proceed directly to the organ from which the chief culture is to be made. If, for instance, it is especially important to obtain a culture from the pulp of the spleen, the mouse is laid on its side with the spleen up, so that it can be exposed without disturbing the other organs too much.

The directions and examples given here may serve as a sufficient introduction to the methods of obtaining cultures from the dead body, as well as to the collection and similar use of fluids and tissues from the living body, where, however, consideration for the patient may necessitate many modifications: *e.g.*, scarification with the heated glass rod must be replaced by a thorough cleansing with disinfectants, the Pasteur pipette must give place to the sterilized Pravaz syringe, etc.

Few appliances are needled for starting such cultures with all requisite precautions, even at the sick-bed or operating table, or for collecting the morbid products in a state of complete purity, for further elaboration in the laboratory. Even when much is aimed at, practically all that is needed is the usual disinfectants (carbolic acid and corrosive sublimate); a little sterile absorbent cotton and filter-paper; scissors, knife, and forceps, wrapped in paper and sterilized at 140° C.; sterilized Pravaz (or Straus) syringe; alcohol lamp; platinum wire; capillary tubes; Pasteur pipettes and others with a capillary constriction (Fig. 53, *a*); small test-tubes containing various culture media, some of them for the culture of anaerobic forms, *e.g.*, with a considerable depth of nutrient gelatin; labels; and cover-glasses. Some of these can be prepared on short notice in the sick-room, some are only needed for exceptional use, while the rest should always be kept in the hospital wards to prevent the waste of much valuable material.

The collection of primarily sterile culture-media, which should really have been considered in the third chapter, is treated here, because the precautions to be observed are exactly the same as in obtaining cultures from the dead body, the collecting apparatus only being different. Usually it is desired to collect the sterile fluids (blood, urine, milk, etc.) either in rather large vessels from which they can be transferred to a number of smaller culture-glasses, or they are at once drawn into the latter.

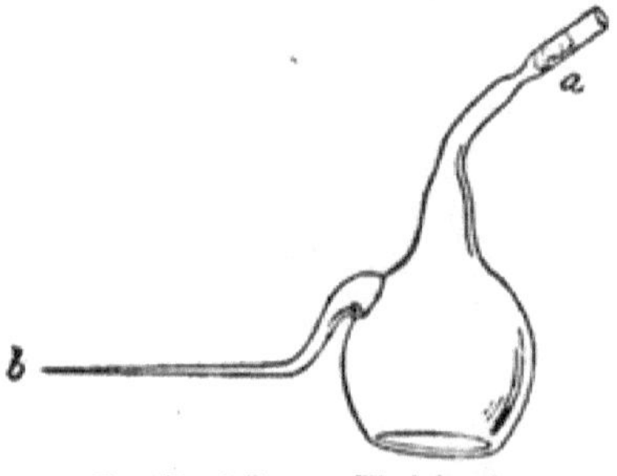

FIG. 67.—A Pasteur Wash-bottle.

In the first case, the so-called Pasteur wash-bottle (Fig. 67) is best used. As the Figure shows, this is really only a very large Pasteur pipette, with a reservoir of peculiar form. When the capillary point *b* has been sealed by fusion, and a cotton plug inserted at *a*, it is sterilized at 150° C. Just before use, the tip *b* is broken off, and the entire capillary part is flamed. Observing the precautions indicated above, the point is inserted, and the receiver sucked full, after which the point *b* is again fused. As soon as possible afterward, the sterile fluid is transferred to smaller culture-vessels, by blowing through *a*, after once more breaking the point off and inverting the reservoir. The test-tubes or Chamberland flasks which have been filled in this way are kept on probation at

30° to 40° C., for several days before being used. In this way it is possible, for instance, to collect large quantities of sterile urine from the newly slaughtered animals at any slaughter-house, by passing the capillary tip of the flask through the wall of the bladder, after the abdominal cavity is opened.

Occasionally it may be convenient to collect the fluid at once in smaller culture vessels. For example, if sterile blood is to be taken from the heart of a newly killed animal, it would coagulate in the larger vessel before it could be transferred to the smaller ones. In such cases, Pasteur pipettes may be used, not closed with simple cotton plugs, as is usual, but with rubber tubes plugged with cotton (Fig. 68), which can be removed and again replaced with far less danger of contamination when the contents of the tube are to be inoculated. The tip of the pipette is opened, flamed, and inserted into the heart of a newly killed healthy animal, the tube sucked full of blood, once more hermetically sealed below, and set aside some time

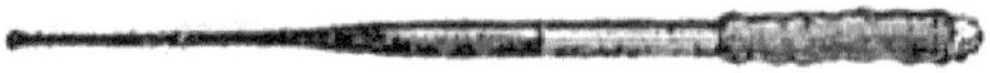

FIG. 68.—Pasteur Pipette with Rubber Cap Plugged with Cotton.

for observation, after which, if free from germs, it may be used directly as a culture vessel. In the same manner, fresh white of egg, containing no germs, can be collected from hens' eggs (*cf.* p. 33).

The collection of sterile fluids from living men or other animals is not always as easy or certain as in these cases, though it is by no means impossible.

For the means of obtaining sterile blood by phlebotomy, see p. 31. Milk can be drawn into sterile test-tubes, after a very thorough cleansing of the surface of udder and teats; but one should expect to find a portion of the tubes accidentally contaminated. Urine can be obtained in a sterile condition either by passing it directly into sterile vessels or drawing it with a catheter, after a careful disinfection of the urethra by washing it out several times.

Now and then very good opportunities occur for the collection of a large quantity of sterile culture fluid from living patients. Dropsical and other similar fluids can be collected free from germs without great difficulty. For this purpose, a small Southey trocar is used, and the fluid is collected in a large flask furnished with a rubber stopper bored with two holes, one of which is plugged with cotton, while the other admits a short glass tube which is joined by rubber tubing to the trocar. The instrument and glass ware are sterilized at 140° C., while the rubber tubing and stopper are steamed before use.

The rules for preparing primarily sterile infusions are implied in the preceding. Meat infusion, for instance, is obtained by cutting a piece of muscle from a recently slaughtered animal, as rapidly as possible, and with all care, and dropping it into sterilized water. Vegetable infusions are generally more easily prepared, if rather large organs, such as fleshy roots and tubers, are employed. The surface of these is cleansed with sublimate solution, after which prismatic pieces are rapidly cut from the interior with a knife that has been heated in the flame so that it hisses with each cut. From the nature of the case, rules for each particular instance cannot be given, often it is necessary to specially prepare the substance, *e.g.*, in forming the primarily sterile infusion of jequirity as described by Salomonsen and Christmas ("Hosp. Tid.," 1884).

It is always to be expected that a certain number of the fluids and infusions prepared in the ways indicated above have become accidentally contaminated in the process of preparation; hence they must never be made use of until they have been kept several days in the thermostat at about 35° C., remaining completely clear and unchanged.

CHAPTER XII.

DISINFECTION EXPERIMENTS.

WHEN it is desired to learn whether a disinfectant is capable of destroying a given contagium, the most natural way is to expose this particular germ, if it is known, or if not, substances which carry it, to the influence of the disinfectant, after which it is used for the inoculation of a suitable organism. But the difficulties in the way of such a test of disinfectants are evident. Often the contagium is one to which our experimental animals are not susceptible, and even when inoculations are possible, they are often too complicated to be feasible on a large scale. Hence, long before pathogenic bacteria could be isolated in cultures, putrefactive bacteria and other micro-organisms were employed as substitutes for the real contagia, in investigations concerning the efficacy of disinfectants, because of the resemblance of pathological processes to those of putrefaction and fermentation.

To-day, with greater reason, non-pathogenic bacteria can be employed as tests of the relative disinfecting power of various substances; for we now know that a large part of the contagia really are bacteria, and a better knowledge of the natural history of the latter has rendered possible the avoidance of numerous errors pertaining to earlier investigations and results. For instance, it is now known that a motile species is not to be regarded as dead because its motions cease; and the extremely different resisting power of different species, and the surprising vitality of spores, are now recognized, as well as the fact that conclusions cannot be drawn for bacteria in general, and hence for all contagia, from what is learned of casual mixtures of bacteria. *E.g.*, the bacillus of typhoid fever is very resistant toward carbolic acid, a circumstance which, according to Chantemesse and Vidal, facilitates their recognition in mixtures of bacteria, since they are capable of developing in nutrient gelatin which contains 0.2 per cent of carbolic acid. On the other hand, substances are to be found which are poisonous to a single species, while they are quite harmless for others: *e.g.*, iodoform, on the antibacterial influence of which, blind reliance was placed before the investigations of Rovsing and Heyn. But while this substance is not fatal to the greater number of bacteria, such as *Micrococcus pyogenes aureus*, the tubercle bacillus (Rovsing) and a number of other pathogenic forms, either within the animal organism or in cultures, it is extraordinarily destructive to the cholera spirillum (Buckner).

Koch's directions for experiments with disinfectants will be followed here very closely, since he has recently stated the problem clearly, and given the simplest and best means of solving it. It is also well known that the researches of Koch, Gaffky, and Loeffler concerning many points have caused an entire revolution in the methods of disinfection.

The rules to be followed in testing the germicidal value of a disinfectant, are, briefly, the following:

Pure cultures of well-known bacteria are used as reagents, not mixtures of unknown composition. For a complete knowledge of the rank of the disinfectant, these pure cultures should represent the several groups of micro-organisms which are active in disease and fermentation, *i.e.*, moulds, yeasts, and bacteria. Bacteria which do not form spores should also be represented, as well as the most resistant bacilli (*e.g.*, the hay bacillus, Chapter IV.). But if it is only desired to learn whether a given substance is capable of destroying all organic life, and consequently all living contagia, it is sufficient to use bacillus spores alone, since these are the most resistant living beings known at the present time.

The forms chosen as reagents ought to be easily recognizable by macroscopic characters; consequently chromogenic species are preferable, such as the black *Aspergillus niger*, the pink yeast [" *Saccharomyces glutinis* "], the blood-red *Micrococcus prodigiosus*, etc., etc. The persistence or absence of ability to develop after exposure to the action of the disinfectant is taken as evidence of their life or death.

According to the nature of the disinfectant, and the circumstances under which it is tested, the cultures employed as reagents are used in various ways. Sometimes they are not prepared in any way; *e.g.*, a tube or flask with the contained bacteria and culture medium can be exposed to a high temperature for a given time, and the power of its contents to develop tested by inoculation in a fresh tube; or a drop of a pure culture can be distributed through a disinfecting fluid and the mixture sown upon a suitable culture medium.

But it is generally more convenient to use the cultures dried upon a solid substance, in which condition they are more easily preserved, transported, and made use of. They may be employed in either of the following ways:

The pure zoöglœa is sliced off from a potato culture by means of a flamed knife, with as little of the tissue of the potato as possible, and laid away to dry in a sterilized glass tray which is wrapped in a double layer of filter-paper instead of being covered by the lid. This prevents the access of dust, while it permits evaporation to go on rapidly enough to completely dry the culture in a couple of days. The dried cultures are kept in sterilized test-tubes plugged with cotton, until they are to be used.

Small pieces of filter-paper sterilized at 150° C. are rubbed upon the potato culture to be used, by the aid of a sterilized glass rod, and dried and preserved as in the preceding case, after being cut into narrow strips with a pair of scissors. These strips are subsequently cut into smaller square pieces when they are to be used. In case of fluid cultures, the paper may be saturated with them, and dried, but it is better to use the next method.

White silk thread is cut into pieces 8 to 10 mm. long, which are sterilized at 150° C. in test-tubes plugged with cotton. Though they may be rubbed upon solid culture, as in the last case, these are especially suited to fluid cultures, into which a large number of them are dropped. After being well shaken about here for some time, they are taken out and dried as above, care being taken to lay them in the glass well separated, since, especially when taken from liquefied gelatin cultures, they are very apt to cling together. They are preserved as in the preceding cases.

Frequent use is made of the spores of *Bacillus anthracis* and *B. subtilis,* in disinfection experiments. To be sure of abundant formation of spores in the former, it must be sown on the surface of nutrient agar, and kept in the brood-oven at

30° to 37° C., for a week or longer. When it is found by microscopic observation that fully developed normal spores are abundant, and the vegetative filaments begin to disappear, 1 to 2 cm. of sterilized water is added to the tube containing the spore bearing culture, which is distributed through it by energetic shaking. The pieces of sterile silk are dropped into the turbid fluid, and preserved in the manner indicated. The hay bacillus is always certain to form its spores in a fluid medium such as bouillon. It is sown in a conical flask containing bouillon to a depth of 2 cm., and kept at 30° to 37° C. A day or two later it has developed a firm spore-bearing membrane upon the surface of the turbid liquid. After some time the culture becomes exhausted and loses its turbidity, the spores sinking to the bottom as a white powder. Three-fourths of the supernatant fluid is poured off, the precipitate of spores is shaken up in the remainder, and silk is saturated with it.

If spores are wanted which have even more vitality than those of these two species, they may be obtained from garden-earth, etc.

The spores of moulds are best prepared in a somewhat different manner. The mould is sown upon gelatinized beer-wort in a pair of small glass trays (Fig. 13). When it has covered the surface and is in fruit, the entire culture can be stripped off in one piece, when it is laid upon a piece of filter-paper in which it is loosely wrapped to prevent the scattering of the spores, and when dry it is clipped into strips.

Garden earth has also, by the advice of Koch, frequently been used as a bacteriological reagent in disinfection experiments. In this case the rule that only pure cultures are to be used is set aside, and in this fact lies the disadvantage of using earth, which contains different bacilli in different places, so that it is impossible to draw general conclusions from the results obtained with a single sample of earth. On the other hand, it has the advantage of being quickly prepared anywhere with ease; but in comparative investigations it is necessary to use small portions of the same bit of earth for the entire series. Pure cultures of the most resistant earth bacilli can easily be obtained by suspending a rather large quantity of dirt in bouillon, which is afterward boiled in a cotton-plugged flask for 15 to 30 minutes, set aside at 30° to 40° C. for a couple of days, and then plated out in agar-gelatin.

It must be remembered that these reagents, excepting, perhaps, the bacillus spores, cannot be kept in a useful state for an unlimited time, but that they generally die after a longer or shorter time. It is, therefore, necessary to ascertain their ability to develop before using them in case they are rather old. It is safest to use freshly prepared material.

The general method of carrying out an investigation is as follows: 1, exposing the bacteria to the action of the disinfectant; 2, removing the latter; 3, sowing or inoculating the exposed bacteria on a suitable culture medium or animal; 4, sowing on inoculating control material at the same time; 5, observing the result of the cultures or inoculations, not merely as to whether or not they grow, but also rapidity of growth, virulence, etc., etc.

When it has been shown by such experiments that a given disinfectant really destroys one of the pathogenic bacteria quickly and surely, it by no means follows that the substance is practically useful for the destruction of this contagium. Before a disinfectant can be recommended, it is particularly necessary to show that it not only kills the germs on our silk threads, bits of papers, and potatoes, but is also capable of attacking them as they occur in nature—in clothing, excrement, sputum, etc. The effectiveness of the most powerful disinfecting agents may under such circumstances be neutralized by chemical changes or the impermeability of the substance which harbors the contagium, which, for example, prevents the successful employment of the corrosive sublimate solution for disinfecting vaults and tuberculous sputum. On the other hand, contagia may occur in nature under circumstances more favorable for their destruction by a given disinfectant than is the case when pure cultures of them are exposed to its action. [Laplace's solution of sublimate in crude hydrochloric acid, diluted for use to the standard ratio of 1:1,000, has the advantage over the simpler aqueous solution, of not so readily forming insoluble precipitates with albuminoids. —W. T.] How to proceed so as to accomplish the purpose of the investigation, is implied in what has been written above. Either the infected excrement, sputum, or clothing is exposed to the action of the disinfectant and then used for inoculation experiments (*e.g.*, by inoculating guinea-pigs with the tuberculous sputum that has been treated with carbolic acid or corrosive sublimate, or causing healthy people to wear pieces of clothing disinfected by heat or chemicals); or the micro-organisms mentioned above are used as reagents, carefully mingled with excrement or sputum, or placed in the pockets or under the lining of clothing in which they are disinfected, and their power of development subsequently tested in the usual way.

Finally, a large number of questions must be answered which are of a clinical or technical nature, etc., and hence lie without the scope of this work (*e.g.*, those concerning the dangerousness of the substance, its possible injuriousness to the objects to be disinfected, offensive odor, expensiveness, etc.); and an opinion as to its practical utility can only be reached after these are considered.

The rules given here must be kept in mind in all disinfection experiments, whether referring to the germicidal power of heat, sunlight (Duclaux), sound waves, electricity, fluids, or gases. But the methods must obviously be modified in their details to suit each case. As examples, a detailed account is given here of the best manner of investigating the disinfecting value of a fluid, and of testing a disinfecting oven.

1. Testing a Fluid Disinfectant.

A. *By the Use of Fluid Cultures.*—Equal quantities of rather dilute and more concentrated solutions of the disinfectant are poured into a number of sterilized vessels. With the same Pasteur pipette, an equal number of drops of a well-shaken pure culture is added to each of these tubes, and carefully mingled with its contents by shaking. The time is noted, and the same number of drops of the culture are then added to a control test-tube containing sterile bouillon or 0.7 per cent solution of table salt.

A large number of test-tubes, with 5 to 10 cc. of nutrient jelly in each are kept at 25° to 30° C., so that their contents remain fluid. From each of the antiseptic solutions that have received bacteria, as well as from the control tube, a drop is transferred to one of the tubes of jelly, the same platinum loop being used for all, to insure the transfer of about the same quantity in each case. This is repeated at longer or shorter intervals for the various degrees of concentration, and the time for which the disinfectant was allowed to act in each case, as well as its concentration, is marked upon the tube.

The appearance or absence of bacteria in these isolation-cultures, as well as their abundance if present, give the data for a determination of the time required by the disinfectant in a certain degree of concentration, to destroy a certain kind of bacteria. It is a weak point in this mode of investigation, that a small quantity of antiseptic is always transferred to the nutrient jelly, together with the bacteria; but this can be assumed to be inactive in the degree of dilution caused by mixing it with the jelly, and this source of error can always be subjected to the test of control experiments.

B. *By Using Cultures Dried on Silk, etc.*—The investigations are carried on along the same lines as the last. Sterile vessels are filled with solutions of an antiseptic of various degrees of concentration. Several pieces of silk, charged with bacteria, are dropped in each. By means of a platinum wire, pieces are fished out at definite intervals, and, after removal of the disinfectant, sown in a gelatinized medium. The removal of the disinfectant is effected by carefully pressing the threads out in sterile filter-paper, washing them in water, and again pressing them between folds of filter-paper. The paper for this use is cut in quadrangular pieces of about 5 sq. cm., folded like sheets of writing paper. These are wrapped in paper in small parcels, and sterilized at 150° C. before use. The control silk is placed in sterile bouillon or 0.7 per cent salt solution, and then washed in the same manner as the pieces that have been exposed to the antiseptic solutions. When the disinfectant has been thoroughly removed by washing and the capillary action of the filter-paper, the pieces of silk are sown in melted jelly or on solid jelly. In the first case, test-tubes are used, which are rolled or laid horizontally while cooling, to facilitate the counting of the germs. In the second case, watch-glasses or small uncovered trays (Fig. 13) are used. Six to ten of these are placed in a larger pair of trays (Fig. 12), sterilized in these at 150° C., and only filled immediately before use. This arrangement facilitates the microscopic control of the cultures, but it makes later contamination possible, and requires a more careful washing than the isolation cultures in test-tubes.

If the threads are not sufficiently washed out, the disinfectant remaining in them may diffuse through the surrounding gelatin in sufficient quantity to prevent all growth about the threads, even when these contain living bacteria—a fact which can cause (and has caused) mistakes. The thoroughness of the washing may be tested in the following manner: Some of the agar-gelatin to be used is inoculated with the same sort of bacteria as those on the threads, just before being poured in the watch-glasses or trays, where it is then allowed to solidify, and the silk is laid on its surface as usual. Some days later, the colonies appear in it, reaching quite to the thread in case the washing has been sufficient, while a sterile zone surrounds the thread if it has not been thorough enough. Attention should be given to this fact in testing the freedom of antiseptic dressings from bacteria by sowing them in nutrient gelatin; control cultures being made by means of a series of thrust-cultures of various bacteria in the immediate vicinity of the pieces of material in question.

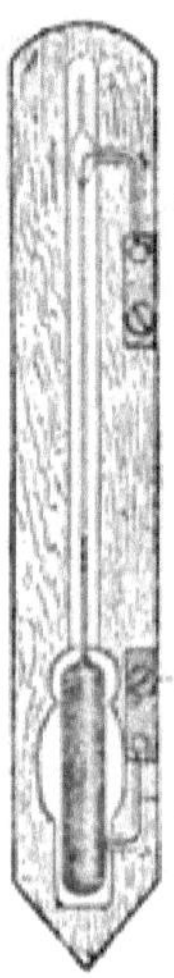

FIG. 69.—Electric Thermometer.

2. Testing a Disinfecting Oven.

For this purpose are needed:

a. A variety of the objects the oven is made for (usually bed-clothes and wearing-apparel) are needed.

b. Thermometers. Usually it is sufficient to have a few maximum thermometers, which, to avoid injury during the investigations, are encased in metal or wood, in the latter case openings being provided to prevent them from checking the penetration of the heat. But if it is wished to learn easily and quickly when a certain temperature (*e.g.,* 100° C.) is reached at a given point within the oven, an electric contact-thermometer Fig. 69) is employed. This consists of a thermometer in the walls of which two platinum threads are fused, one (p_1) entering the mercury in the bulb, the other (p) touching the mercury only when the boiling-point of water (or other desired temperature) is reached. By means of screws (s and s_1), the thermometer is connected with wires that may be passed through the door of the oven and inserted in an electric circuit including a bell which sounds when the circuit is closed.

c. Bacteria.—The principal of these are resistant spores, like those of *Bacillus anthracis, B. subtilis, Tyrothrix scaber* (Duclaux), and certain earth-bacilli. The material is prepared as indicated on pages 123–125. For packages, filter-paper is used, folded like the papers used by druggists for "powders," and sterilized at 150° C. The papers are marked and numbered with a lead-pencil, and those containing the different sorts of spores are collected in small parcels which can eventually be tied to a thermometer.

When the material for the test is ready, the oven is packed as it is intended to be in regular use. The thermometers and test-bacteria are distributed through the oven during this process, not merely between and upon the clothing, but within it, in places where contagia might easily exist, but where the heat penetrates only with difficulty. For instance, mattresses and pillows are ripped open enough to allow of the insertion of thermometer and bacteria into the straw, hair, moss or feathers, and again closed by sewing or the use of safety pins; one or more blankets are rolled tightly about thermometer and bacteria, etc. If the bacteria are destroyed under these circumstances, it is evident *a fortiori* that as good results are attainable with the looser packing which may be employed in the daily use of the oven. Care must be taken that the thermometers are not all collected about a single place, but are distributed through the oven.

After the oven is packed, it is best to make a diagram (Fig. 70) of its contents, that the temperature reached and its influence on the bacteria may be subsequently noted at various points, giving an excellent synopsis of the results of the test.

At the end of the experiment, as accurate notes as possible are made of the time for which it was continued, the amount of fuel consumed, etc.; the articles are unpacked, and their appearance, degree of dryness and brittleness, etc., observed, and the thermometer readings made.

The packets of bacteria are wrapped in paper, in which they can be carried to the laboratory without danger of contamination. Cultures are made from them, as outlined on p. 127, either in test-tubes or trays. But it is to be observed that the latter sort of cultures is not suited to garden-earth, because, even when the dirt is poured from the papers very slowly, particles easily fall into adjacent trays. It must, therefore, be poured on the surface of gelatin which has been allowed to harden obliquely in test-tubes. In larger series of comparative tests, it is well to place the cultures in a brood-oven at 20° C., but isolated experiments are as well made at the varying temperature of the room. Most cultures require to be kept only a week, but those from garden-earth must be watched for a couple of weeks, as they occasionally contain resistant forms which develop very late. It is necessary to note not only whether or not growth occurs, but also whether there is any retardation of the development of the bacteria, and, in isolation cultures, their number.

The final results of the experiment are most conveniently

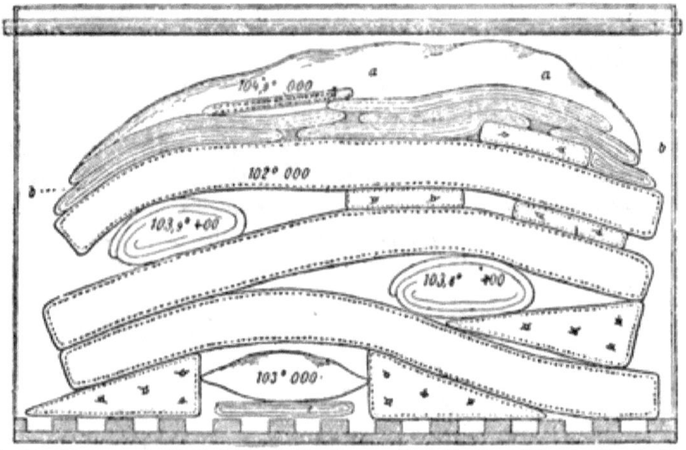

Fig. 70.—Diagram of a Packed Disinfection Oven, with Indication of the Results of a Test.

registered by noting the thermometric and bacteriological observations upon the diagram representing the contents of the oven. In Figure 70, + indicates growth; 0 no development; and the species of bacteria are indicated by the relative position of the marks, which are always arranged in the same sequence—earth, hay, anthrax-bacilli. *E.g.*, within the upper mattress of the Figure, the temperature reached was 102° C., and all three species of bacteria were killed, while the earth-bacilli survived the disinfection within the rolled woollen blankets, where the temperature was between 103° and 104° C., etc.

As has been intimated above, one should be careful not to

confound the power of a disinfectant to check or prevent the growth of the bacteria, with its power of killing them. That a substance, added in a certain quantity to a readily putrefying fluid, prevents decomposition, by no means requires that it should have killed the bacteria in the fluid. The development and dissemination of the germs may, perhaps, have been prevented; and it may be that when brought into a new and favorable soil they would prove very well able to develop. It is not superfluous to mention this perfectly obvious fact, because it is often overlooked, giving origin again and again to errors.

If it is wished to learn the degree of concentration in which a given material begins to check the growth of different micro-organisms, and that in which it renders growth entirely impossible, the following method is employed, bacteria and moulds being used, selected as indicated on p. 122.

A suitable number of test-tubes are filled with nutrient material (generally gelatin) adapted to the organisms selected. Some of these glasses are left untouched, serving for control, while larger or smaller quantities of the substance to be tested are added to the rest, so that all degrees of concentration are obtained. In preliminary experiments, it is best to pour nearly equal quantities of gelatin into all of the tubes, adding to them different quantities of a strong solution of the substance, of known concentration, by means of glass pipettes graduated to 0.05 cc. From these data, the percentage of the substance in each culture glass can readily be calculated.

The species of bacteria on which it is desired to test the disinfectant is sown in the vessels prepared in this way, the material used for inoculating all being taken from a single culture and by means of the same needle or pipette. The quantity used for inoculation, and the manner in which the needle is used should also be, as far as possible, exactly the same for the different glasses, which are then observed daily, the least degree of concentration which permits no growth along the thrust and the least which perceptibly checks the growth of the bacteria, being noted. The latter point is chiefly indicated by a retardation of development, and by a change in the macroscopic habit of the colonies.

Figure 71 illustrates this. It represents three test-tubes with peptonized meat-infusion gelatin to which 0.05 per cent of carbolic acid was added in the first, 0.4 per cent in the second, and 0.7 per cent in the third. The same bacillus was sown in all three glasses, which eight days later presented the appearance represented. In the first tube, the appearance of the culture was the same as in the untreated control-tube—a large close irregular felt of bacilli floating on a quantity of melted gelatin at the very top, and below, above the thrust in the still solid gelatin, a number of dot-like colonies, from some of which, especially the uppermost, fine undulating threads radiate in all directions. In the second tube, during the same time, the felt of bacilli next the surface had not reached the glass, liquefaction was limited to a small hemispherical depression immediately about the needle-thrust, while no trace was to be seen of filaments from the deeper colonies, which did not approach the bottom so nearly as in the first glass. The third tube showed little surface growth, no liquefaction, and only a short, firm, inversely conical colony in the solid gelatin. Very similar differences are found in cultures of Koch's bacillus of mouse septicæmia, grown in gelatin with and without carbolic acid. The peculiar cloud-like fine growth never appears in the cultures containing much carbolic acid, in which the bacilli show as small, dense, dot-like colonies.

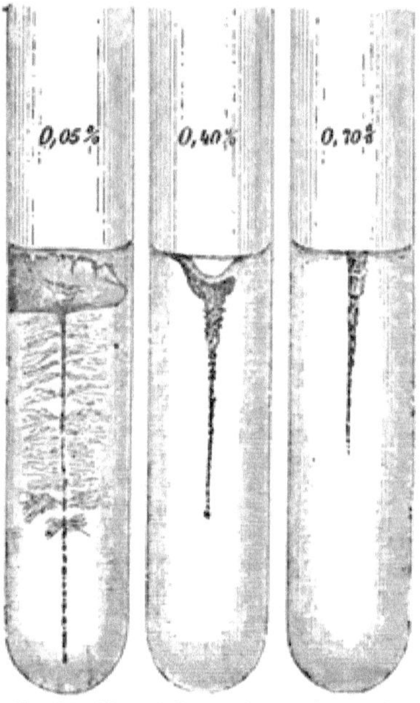

FIG. 71.—Three Cultures of *B. anthracis* after Seven Days' Growth in Peptonized Meat Gelatine, to which has been Added the Percentage of Carbolic Acid Indicated in Each.

This prepares the observer for morphological changes in the microscopic appearance of the bacteria so grown; and it must also be remembered that under such conditions a permanent physiological transformation of the bacteria has been induced, as in the attenuation of the virulence of *B. anthracis* by cultivating it in carbolic acid (Toussaint, Chamberland, and Roux)—a result which gives especial interest to experiments of this sort.

CHAPTER XIII.

MICROSCOPIC EXAMINATION, AND STAINING OF BACTERIA.

One side of the microscopic examination of bacteria has already been considered in Chapter IX. (moist chambers), and in sketching the plate-cultures of Koch, where not only the peculiarities of the colonies (whether entire, lobed, fringed, smooth, granular, wrinkled, etc.) can be observed with weaker and medium powers, but which without further contrivances, may also be studied with higher powers, the most superficial colonies being even accessible to the most powerful immersion lenses when covered with a cover-glass.

If it is wished to study individual cells more closely, the colony need only be touched with a sterile platinum needle, and the adhering material distributed through a small drop of 0.7-per-cent solution of table salt (free from bacteria), spread in a thin layer under a cover-glass. If the bacteria grow in a fluid medium, a small drop of the culture is placed under the cover-glass without further treatment except for final dilution with the standard salt solution. The addition of such a neutral fluid may occasionally facilitate the investigation, not only by isolating the germs, but by modifying the differences of refraction.

If such preparations are to be examined for a longer time or with an immersion lens, it should be sealed with paraffin, which is best applied by means of a glass tube 4 mm. in diameter filled with paraffin and drawn out to a short open point, from which the melted paraffin flows.

The movements of living bacteria cannot be observed long in such sealed preparations, since they are stopped by want of oxygen. This may be put off for a time by having a number of rather large air-bubbles under the cover-glass, or, in case the bacteria are contained in water, by introducing one of the filamentous green algæ (Engelmann). Impelled by their need of oxygen, the bacteria collect about the bubbles or algæ (the latter of which continue to set free oxygen, as a result of assimilation), and keep up their motions until the supply is exhausted or the activity of the algæ ceases. The study of the manner in which the bacteria move can be facilitated by slightly coloring them, as indicated below. The movements of the very active species can be checked by compressing them strongly by pressure on the cover-glass. All motion in a limited part of the preparation may be stopped by the same means, individuals here and there afterward recommencing their movement, but at first very slowly.

The best way of studying the movements of bacteria is, naturally, by the use of a moist chamber, Ranvier's moist chamber, and the method indicated on p. 98 is especially advisable, as the difficulties of using the hanging drop are thus avoided.

The use of staining fluids, especially certain aniline colors, is an indispensible adjunct in the microscopic study of bacteria. The employment of the methods of staining now used, not only greatly facilitates the demonstration and observation of bacteria in fluids, but the detection of structural peculiarities which escape observation in the unstained cells, *e.g.*, the flagella of certain bacteria, the peculiar ends of the cells of *B. anthracis*, etc.; and it also discloses chemical differences which are sometimes of extraordinary importance for clinical diagnosis, *e.g.*, in demonstrating the tubercle bacillus, staining has also rendered it possible to obtain handsome permanent preparations of free bacteria, and, above all, to demonstrate bacteria within the tissues and to study their distribution in the organs and their relations to the cells. The development of the technology of staining to the high point it has now reached is due chiefly to three men, Weigert, Koch, and Ehrlich.

The dyes used in bacteriological investigation nearly all belong to the large class of aniline colors, and especially to the group called basic by Ehrlich, who first called attention to the fact that the aniline colors are divisible into two main groups: the acid colors, in which the staining principle is an acid (*e.g.*, ammonium picrate); and the basic, consisting of a staining base in combination with an acid that does not stain (*e.g.*, acetate of rosaniline). Methylene blue, fuchsin, gentian-violet, vesuvin, etc., belong to this group of basic dyes (used by histologists as nuclear stains), which have proved especially adapted to the staining of bacteria, free or *in situ*, so that it has proved possible by their use to stain all known pathogenic bacteria in sections, strongly, permanently, sparingly, and distinct from the invaded tissue. Ehrlich likewise calls those dyes neutral which are formed by the union of a base and acid both of which are capable of staining (*e.g.*, picrate of rosaniline).

Naturally, it is not always possible to at once secure all of these desirable results equally well, and especially soft staining of the histological elements is often advantageously made to give place to the intense coloring of all bacteria present. But Baumgarten has, for instance, succeeded by a rather complicated method in showing upon the same slide the tubercle-bacilli and the caryokineses resulting from their invasion. Baumgarten's method is the following: The diseased rabbit is killed, and small pieces of the tuberculous organs are cut out as quickly as possible and at once cast into 0.2-per-cent chromic acid, where they are hardened for forty-eight hours, then washed out thoroughly in running water for as much as twenty-four hours, and rehardened for twenty-four hours in strictly absolute alcohol. The thinnest possible sections are laid in freshly prepared aniline-methyl violet (No. 8) for a long time (as much as forty-eight hours), washed not to exceed thirty seconds in 1 part of nitric acid to 5 of water, the decolorization being then completed in 60-per-cent alcohol (No. 17), after which they go into a mixture of equal parts of a concentrated alcoholic solution of fuchsin (No. 1) and distilled water for half an hour to an hour, then for five to ten seconds in an aqueous solution of methylene blue (1 : 1,000), and finally for five to ten minutes in absolute alcohol, changed once or twice. The sections are finally mounted in balsam thinned with bergamot oil, but without chloroform.

Instead of the more complicated method of staining the nuclei with fuchsin and methylene blue, according to Baumgarten a concentrated solution of vesuvin in 1-per-cent acetic acid may be used.

For securing intense and permanent staining, the same means are employed as in dyeing on a larger scale; *i.e.*, by the prolonged action of the staining fluid; warming the latter, either for a long time in the thermostat at 40° to 50° C., or more intensely but for a shorter time over the flame; or by the use of mordants, *e.g.*,—potash, carbolic acid, aniline oil, or tannin—which have a certain tendency to unite with both the dye and the object to be stained, so that they serve as a sort of connecting link.

The possibility of staining bacteria in contrast with the surrounding tissue is partly due to the different "elective" power of various dyes (Ehrlich), *i.e.*, their different power of staining certain tissue elements; and partly to the different strength with which they stain various parts of the preparation, *i.e.*, the different stability of the union they form with them. The following experiment of Ehrlich and Schwarze illustrates in a striking way the different elective power of three acid aniline colors, aurantia, indulin, and eosin. When a cover-glass preparation of blood is made by distributing it in a thin layer, drying it, and heating to 120° C., and the pigments are tested on it, it is seen that each of the latter possesses the power of coloring the red corpuscles as well as the nuclei of the white corpuscles and the peculiar granules, found in some of the white cells, which Ehrlich calls α-granules. But if the blood-preparation is treated with a solution of the three dyes in glycerin (made by adding an excess of eosin and indulin to a mixture of 1 part of saturated solution of aurantia in glycerin, and 2 parts of pure glycerin), the hæmoglobin is colored yellow by the aurantia; the nuclei, gray or black by the indulin, and the α-granules, red by the eosin. The fact that certain parts of the preparation hold the coloring matter more tenaciously than others, is used in demonstrating bacteria in sections, the general plan being to overstain the entire preparation, subsequently securing the differentiation by decolorizing (and finally staining in some contrasting color) certain parts, leaving the others colored as at first. The degree of decolorization can be regulated by the substance chosen to effect it. If, for example, we have a section of an organ containing bacteria, which has lain in a strong solution of methyl violet, and wash it in water, microscopic examination will show the bacteria, as well as the protoplasm and nuclei of the cells, colored deep violet (diffuse staining). If it is washed in dilute acetic acid (p. 141, No. 11), the protoplasm becomes decolorized, while the bacteria and nuclei retain the color (nuclear staining). If it is washed in potassium carbonate (No. 15), the coloring matter is also driven from the nuclei, the bacteria alone remaining stained (isolated staining of the bacteria). By still more powerful agents, such as 25-per-cent nitric acid or hydrochloric acid, most species of bacteria can likewise be decolorized, so that only a very few sorts remain stained (tubercle stain); but repeated treatment with strong solutions of mineral acids, if continued sufficiently long, also removes the stain from these species, and it must be remembered throughout that the time for which the preparations are exposed to the action of the decolorizing agent is of decisive importance for the result, since the power of the different elements to resist decolorization usually differs only in degree, not in kind. In each case, therefore, it is necessary to avoid decolorizing too little or too much, and as it is not easy to give very exact instructions as to the time needed, this is one of the points on which long experience plays an important part.

To further differentiate the bacteria from the tissue elements, and to bring the form and disposition of the latter out more clearly, multiple staining (usually double staining) can be employed, either by using a new staining fluid on the partly decolorized preparations (successive staining, *e.g.*, No. 17), or by treating them with a fluid which at once decolorizes and restains them in part (replacement staining). Treating the same preparation with two colors is also sometimes resorted to for another purpose, namely to render the staining more resistant toward decolorizing substances, either by forming a new compound color, or otherwise, *e.g.*, the gentian-violet-iodine method (Gram) and the fuchsin-methylene-blue method described on p. 147.

Occasionally double staining may be secured by the employment of a single dye, *e.g.*, when methylene blue is used to stain a section which includes bacteria as well as the peculiarly granulated connective-tissue cells known as "Mastzellen," the bacteria are colored blue, and the granules violet. These granules (the γ-granules of Ehrlich) also deserve notice for another reason. Since, like bacteria and nuclei, they are stained by basic aniline colors, and have about the size of micrococci, they may be mistaken for the latter, as has often been the case with inexperienced observers. As an especially favorable object for the comparative study of micrococcus cells and these γ-granules, I can recommend the mesentery of a rather lean mouse, dead of suppurative peritonitis as a result of inoculation with pyogenic staphylococci. The mesentery is spread out on a cover-glass which is slipped beneath it while it is still attached to the intestine, by the weight of which it is stretched over the glass. As soon as it has dried fast, the part projecting around the edge of the glass is cut away, and it is treated as if it were an ordinary cover-glass preparation (by drying, passing through the flame, etc.), and stained by methylene blue. Among the blue cocci will be found small clusters of granules of unequal size, grouped irregularly around an unstained nucleus—for the nuclei of these cells, unlike those of other connective-tissue cells, are not colored by this mode of staining with basic aniline colors. In double-staining, colors are naturally chosen which contrast sharply and prettily with one another—*e.g.*, fuchsin and methyl green; fuchsin and methyl blue; methyl violet and vesuvin [or methyl violet and eosin].

An extremely large number of dyes have gradually come into use in various ways for staining bacteria; but only a limited number of stains and methods will be considered here, and it is best, for beginners especially, to be content with one or two of the most universal and important methods, practising them carefully until they are mastered, before going further. It is not possible to limit one's self exclusively to a single dye, for the reason that the behavior of different bacteria toward the various staining fluids, and when different methods are used, is of diagnostic importance, while there is no one dye which stains all bacteria equally well. *E.g.*, the slight power of methylene blue to stain the bacillus of leprosy can be used to distinguish the latter from that of tubercle, which otherwise resembles it closely (*cf.* p. 155). In other cases, the behavior when the Gram method is employed can be utilized for diagnosis; *e. g.*, with gonococci, which (like the typhoid bacillus, the cholera spirillum, and the microbes of chicken cholera and the septicæmia of mice) are decolorized when the Gram method is used, while other pyogenic micrococci, which closely resemble them, retain the stain. The most nearly universal pigment is methylene blue, a substance introduced by Ehrlich, which can well be used as the principal stain. By employing it in the manner indicated below (No. XIII.), Kühne succeeded in staining *in situ* all of the bacteria that he investigated. In leprosy and mouse septicæmia, alone, he failed to obtain completely satisfactory results, and for the bacilli of these diseases he employed fuchsin. Consequently, it is best to carefully practise this "universal method with methylene blue;" and if to this are joined one of the tubercle methods (*e.g.*, No. XVIII.), and the Gram method, one is well equipped, so far as sections are concerned. It is also best to confine one's self to a few methods for cover-glass preparations.

The materials employed, in the methods described below, are the following: Distilled water, absolute alcohol, glycerin, acetic acid, hydrochloric acid, nitric acid, sulphuric acid, chromic acid, carbolic acid, aniline oil, potash, potassium carbonate, potassium acetate, potassium iodide, lithium carbonate, iodine, clove oil, bergamot oil, cedar oil, turpentine, xylol, Canada balsam, shellac, paraffin, fuchsin, methylene blue, methyl violet, gentian violet, methyl green, vesuvin, picro-carmine, and extract of logwood. Different aniline colors bearing the same name may differ to such an extent as not to be equally adapted to histological purposes, so that attention should be given to the source and trade mark of good sorts.

All of the aniline colors named should be kept in stock in the dry form. Fuchsin, methylene blue, methyl violet, and gentian violet may be further kept in a saturated alcohol solution (No. 1), *i.e.*, 25 gm. of the dye to 100 gm. of absolute alcohol, which always leaves an abundant excess of undissolved pigment in the bottom of the flask. Vesuvin (or Bismarck brown) is best kept only in powder, but if a solution is to be kept, this is best made (No. 2) by saturating equal parts of water and glycerin.

For use, aqueous solutions of the powders (No. 3) may be directly prepared, but these do not keep long, and are therefore always freshly prepared from water free from bacteria, and are filtered before use. It is very convenient to keep a little of the dry pigment constantly used, in this way, on a little filter in a glass funnel. When it is needed a small quantity of water is poured over it, and the filtrate is collected, the powder soon drying again. Dust is kept from it by covering the funnel with a layer of filter-paper.

When it is not exceptionally necessary to avoid every trace

of alcohol in the staining fluid (as is, for example, the case when staining living bacteria), the aqueous solution is replaced without disadvantage by that (No. 4) made by adding a suitable quantity of the saturated alcoholic solution (No. 1) to water. This diluted alcoholic solution is more durable than the aqueous, but it usually requires renewal once or twice a month, so that it is best to prepare it when needed, by adding five to six drops of No. 1 to a watch-glass full of distilled water.

The principal mordants used are 0.01-per-cent solution of potash (Koch, Loeffler); 5-per-cent carbolic acid (Ziehl); and aniline-water (No. 5), a concentrated aqueous solution of aniline oil, prepared by very thoroughly shaking about 1 part of aniline oil and 20 parts of distilled water in a test-tube, allowing it to stand five minutes, and filtering through a filter moistened with distilled water. (It must be perfectly free from turbidity, or it should be again shaken, and refiltered.) The staining fluids with mordants for which we shall find application, are:

(No. 6.) Kuehne's carbolic blue. 1.5 parts of methylene blue, and 10 parts of absolute alcohol are triturated lightly in a watch-glass with 100 parts of 5-per-cent carbolic acid which is added little by little. When all is dissolved, it is bottled. To facilitate rapid preparation, several test-tubes with feet may be graduated to 20, 22, and 24.2 cc. (*cf.* Fig. 72).

(No. 7.) Ziehl's carbolic fuchsin. 1 part of fuchsin, 10 of alcohol, and 100 of 5-per-cent carbolic acid.

(No. 8.) Aniline methyl violet (Ehrlich-Weigert). 11 cc. of the saturated alcoholic solution of methyl violet, 10 cc. of absolute alcohol, and 100 cc. of aniline-water. Or the dry powder may be added in excess to aniline water. [In Koch's laboratory it is customary to add the stock alcoholic solution (No. 1) to a watch-glass of aniline-water until the latter is shown to be saturated by the formation of a film at top.]

(No. 9.) Aniline gentian violet (Ehrlich). 5 cc. of the saturated alcoholic solution of gentian violet, to 100 cc. of aniline-water.

(No. 10.) Loeffler's alkaline blue. 30 cc. of the saturated alcoholic solution of methylene blue to 100 cc. of 0.01-per-cent caustic-potash solution.

A large part of the chemicals enumerated above are used for washing preparations for decolorization or differentiation.

For this purpose, water, alcohol, and glycerin (all perfectly free from acid) are used, as well as clove-oil and aniline-oil, and acids and salts in the following forms:

(No. 11.) Very dilute acetic acid (0.5 per cent to 1 per cent).

(No. 12.) Very dilute hydrochloric acid (10 drops of the acid to 500 gm. water).

(No. 13.) 75 parts of water containing 25 parts of nitric acid (Ehrlich), or the same quantity of hydrochloric or sulphuric acid.

(No. 14.) Lithium water (Kuehne). 6 to 8 drops of a concentrated aqueous solution of lithium carbonate and 10 gm. water. For neutralizing an acid washing fluid.

(No. 15.) Potassium carbonate solution (Koch). Equal parts of a saturated aqueous solution, and water (Koch); or 2 parts of a 2-per-cent aqueous solution, and 1 part of absolute alcohol (Malassez and Vignal).

(No. 16.) Gram solution. Iodine 1 part; potassic iodide, 2 parts; distilled water, 300 parts.

(No. 17.) Alcohol, 60 parts; water, 40 parts (Koch). For colored wash-alcohol (Kuehne), see p. 151.

(No. 18.) Aniline-oil blue. Kuehne recommends rubbing as much methylene blue as can be raised on the point of a knife, with 10 gm. clarified aniline oil, and allowing the excess of undissolved pigment to settle in a bottle. A few drops are added to aniline oil in a watch-glass, until the desired concentration is reached.

The manner of using these staining and decolorizing agents, as well as the others for hardening, anhydrating, clearing, mounting, and sealing preparations, receives more detailed consideration under the several methods described below.

In what follows, it is assumed that the common microscopic appliances and methods are understood. Staining is usually effected in small glass trays (Fig. 13), "individual salt cellars" or watch-glasses. Some of the trays must have ground tops so that they can be tightly sealed with a glass plate if they are to kept at an elevated temperature for some time. Watch-glasses, which are inconvenient because of their lack of stability, are necessary if sections are to be stained by heating over the flame. In this case, they are conveniently placed upon the stand shown in Figure 72, which consists simply of a strip of sheet tin with three holes having a somewhat smaller diameter than the watch-glasses. The stand is also useful for filtering, as indicated in the cut.

Staining Bacteria in Fluids.

A. *By the Simple Addition of the Staining Fluid.*—A little drop of very dilute aqueous solution (No. 3) of fuchsin or methyl green (Macé) is placed on a slide, a small quantity of the culture, etc., is distributed through it with a platinum needle, and a cover-glass applied. The staining fluid must usually be so dilute that it appears nearly colorless under the microscope. When pure and extremely dilute aqueous solu-

Fig. 72.—Tin Support for Funnels and Watch-glasses, in Staining and Filtering.

tions of fuchsin are used, the bacteria can remain alive and continue their motions after staining (Salomonsen).

If a drop of the culture and one of a more concentrated staining fluid than in the last case are placed in proximity on the slide and covered with a single cover-glass, a series of bacteria may be found in the same preparation, showing all gradations from cells that are unstained to others strongly overstained.

A rapid microscopic view of a large number of crowded colonies in a plate-culture is obtained by the following method: A well-cleansed cover-glass is laid upon the gelatin over the colonies and pressed firmly into contact with it everywhere, by a glass rod or pair of forceps. Fragments of the superficial colonies adhere to the cover, forming an "impression-preparation" of the culture, so that when it is removed from the gelatin and lowered on to a drop of staining fluid on a slide, the opportunity is given to see small samples of the contents of all of the colonies touched. Any of these preparations can be sealed with paraffin, as described above.

B. *Staining Preparations Dried on the Cover-glass.*—Weigert was the first, in 1876, to recommend an aniline color (methyl violet) for staining bacteria. Shortly afterward, without knowing of Weigert's work, I introduced fuchsin as particularly good for staining bacteria, and indicated especially its value as a means of diagnosis in the microscopic analysis of putrefying blood. By the introduction of the basic aniline colors, the demonstration and examination of bacteria in fluids was greatly facilitated; but a further and decisive advance in methods was made in 1878, when Koch showed how bacteria might be colored after drying them in a very thin layer upon the cover-glass, which, likewise, rendered possible the complex micro-chemical treatment of free bacteria which now plays so important a part. The great advantages of this method have caused it to be almost exclusively used in all chemical examinations of blood, pus, urine, and sputum; consequently the making and examining stained cover-glass preparations will be sketched in detail in the following pages. Obviously, the use of such preparations does not render superfluous the simple staining already described, since the latter completely preserves the form and turgescence of the cells, while a shrinking always results from drying them.

New cover-glasses are carefully washed in warm water, dried, and treated with absolute alcohol for the removal of all grease. Those that have been used are first laid in strong mineral acid (hydrochloric or sulphuric), then washed clear in water and rinsed until every trace of acid is removed, when they are treated as if new.

The fluid is best spread in a thin layer, by placing a small drop on a cover-glass, laying another cover upon it, and separating them by drawing one over the other, in which way two preparations are obtained. It is also possible to scrape the fluid in as thin a layer as possible by the edge of a second cover-glass. The covers are then laid to dry under a bell-glass, with the smeared side up. If the drying is to be accelerated by warming, this must be effected at a very low temperature. The air-drying must be complete before the next step is taken.

When such a dried preparation of blood, pus, sputum, or other fluid which contains albumen, is at once stained, two defects are sometimes observed; sometimes the dried film separates in part from the glass; sometimes the presence of soluble albuminoids causes disturbing precipitates in the preparation. These are avoided when the albuminoids are rendered insoluble by suitable hardening. The second defect can also be avoided by using aniline brown (Bismarck brown or vesuvin) in glycerin (No. 2), and washing in pure glycerin (Koch). Absolute alcohol (Koch) may be used for this purpose, the cover-glass being placed in it for some time, but the requisite time varies much for different preparations. Partly for this reason, and partly because of its slowness, this method is far inferior to that of Ehrlich, by heating up to 120° to 130° C. for two to ten minutes. (In a protracted exposure to this temperature, *e.g.*, for an hour, as in Ehrlich's studies of the granules of the white blood-cells, the bacteria have been found to lose their power of staining.) This heating may be effected in the sterilizing oven (Fig. 1), but it is more convenient to follow Koch in passing the cover three times through the flame of a Bunsen burner. The cover-glass is seized with the forceps by one corner, the smeared side up, and passed three times through a vertical circle about a foot in diameter, the clean side of the cover being brought down against the top of the flame each time. The best rapidity for the flaming, upon which success depends, is reached by taking about three seconds for describing the three circles.

Of the various ways in which the staining fluid and the hardened preparation can be brought together, we usually employ that of allowing the cover-glass to float upon the fluid, film downward. In all of the following manipulations, care must be taken to remember which side of the cover bears the bacteria, for when the film is very thin it may be difficult to distinguish it. Sometimes it may be recognized by scratching upon the cover with a sharp pin, the irregularities of the smeared side being felt, or the scratches seen.

The smeared and dried cover-glasses can be kept, with or without hardening, for an indefinite time before staining, being simply laid in an ordinary cover-glass box, the different sets separated by labels of the size of the covers. [For clinical purposes, it is convenient to use a small oblong box, a couple of inches long, with a rack inside like a slide-box. Twenty or more clean covers are easily carried in such a box, which is small enough to be slipped into the vest pocket or instrument-case, so that they are always ready for use at the desk or bed-side.—W. T.] Two opposite corners are lifted with the thumb and forefinger of the right hand, the cover is held horizontal, and allowed to fall upon the fluid from a height of several centimetres. The time required for the action of the staining fluid varies according to circumstances from a few minutes to an entire day. Long staining at the temperature of the air can often be replaced by a shorter staining at a higher temperature (Koch, Loeffler), *e.g.*, No. IV., p. 147, *infra.* This is best effected by setting the watch-glass of staining fluid on which the cover floats, upon the tin stand (Fig. 72), and heating it with a small flame until vapor rises freely from the surface. The lamp is then removed for a short time, after which the heating is renewed, and this is repeated several times for a few minutes, or longer, as may be necessary. Sometimes it may be convenient to heat the fluid until it begins to boil, *e.g.*, in No. IV.

When the staining, or, rather, overstaining of the preparation is finished, the latter is washed (partly decolorized or "differentiated"). This is effected according to circumstances by the use of one or other of the fluids enumerated above (*cf.* pp. 140 and 141), which is then removed with water either by rinsing under the faucet or with a wash-bottle, or by moving the cover back and forth in a large dish of distilled water. Washing out the excess of color is the most difficult part of staining, because no rules can be given as to the length of time it ought to be continued in each case: frequently the right time is only learned by experiment. When it has been finished, the preparation can be examined at once in distilled water or glycerin.

It should also be observed that preparations stained with Bismarck brown or vesuvin (No. 2), can be permanently mounted in glycerin, while this substance decolorizes preparations stained with the other aniline colors. On the other hand, a concentrated solution of acetate of potassium (1 : 2) serves especially well, according to Koch, for the preservation of bacteria stained with violet or fuchsin, and, it may be added, of unstained bacteria; but it should not be used for those stained brown.

If the preparation is to be permanently mounted in balsam, it is next dried, by carefully wiping off the clean side of the cover with a piece of soft linen, and holding it obliquely with one edge upon a piece of filter-paper, so that as much water as possible will flow off or be absorbed by the paper. Any large drops remaining on the film are carefully removed with filter-paper, and the cover is allowed to become air-dry. To hasten the drying, the cover may be pressed between folds of filter-paper (Ehrlich), waved back and forth through the air, or air may be blown over it by means of a rubber bulb furnished with a pointed glass tube (Kuehne).

The dried preparation is mounted in Canada balsam. It is most convenient to use this greatly thinned with xylol, and kept at hand in collapsible tubes. Turpentine can also be used for this purpose, but chloroform [and benzol] are to be avoided, as they remove the basic aniline colors. Of the ethereal oils, clove-oil is especially apt to decolorize the bacteria, so that sections should rather be cleared with turpentine, oil of bergamot, or cedar-oil. A drop of this is applied to the middle of the slide, and the cover inverted upon this, spreading it in a thin layer by its weight. Further treatment is unnecessary, but as it requires a long time for the xylol-balsam to harden sufficiently to fix the cover immovably, it is convenient to seal the preparation by means of a filtered alcoholic solution of shellac, which may be given a handsome green color by the addition of a little methyl green. Several layers of this cement are painted around the edge of the cover.

In the preliminary examination of a fluid containing bacteria, a less complicated method is used. After drying and flaming the cover-glass preparation, it is stained by a diluted alcoholic solution (No. 3) of methylene blue or fuchsin, rinsed with water, and examined at once in water, or, after drying, in cedar-oil (*cf.* No. 1).

Methods of Staining Cover-Glass Preparations.

(I.) *Koch's Original Method* (1878).—The dried film is stained for a few seconds, or a little longer, with a solution of methyl violet or fuchsin (a few drops of the saturated alcoholic solution to 15 to 20 cc. water), washed with water or acetate of potassium (1 : 10), dried, and mounted in balsam;

or it is stained with vesuvin dissolved in equal parts of glycerin and water, and washed and mounted in glycerin.

(II.) *Kuehne's Methylene-Blue Method.*—The film, after drying and flaming, is stained for five minutes in carbolic blue (No. 6), rinsed in water, differentiated by laying in dilute hydrochloric acid (No. 12) for a few seconds or a little longer according to the thickness of the film, rinsed a moment in lithium-water (No. 14), washed under the water-jet for fifteen seconds, dried—if necessary by use of the bulb—slightly warmed over the flame, cleared in xylol, and mounted in balsam.

(III.) *Gram's Method.*—The flamed film is treated for one to three minutes with aniline gentian violet (No. 9), laid in the iodine solution (No. 16) for an equal time, washed in alcohol till it appears completely decolorized, dried, and mounted in balsam.

(IV.) *Koch-Ehrlich-Weigert Tubercle Method.*—The dried and flamed film is heated in a watch-glass of Weigert's aniline methyl violet (No. 8) until bubbles begin to form beneath the cover-glass, when it is allowed to stand five minutes, decolorized in a tray of 25-per-cent nitric acid, in which it is moved back and forth for at most five seconds, immediately rinsed in 60-per-cent alcohol (No. 17) until the blue color disappears (usually not over one or two seconds), counter-stained for five minutes in a saturated aqueous solution of vesuvin, rinsed in water, dried, and mounted in balsam.

(V.) *Ziehl-Neelsen Tubercle Method.*—The flamed film is floated on a watch-glass of carbolic fuchsin (No. 7) three to eight minutes (with heat), decolorized in 25-per-cent nitric or sulphuric acid, treated with 60-per-cent alcohol until only a rosy tint remains, the acid then being completely washed out in a large quantity of water, dried, and mounted in Canada balsam, with or without previous clearing in xylol.

[For rapidly staining the hardened sputum film, the following commonly-used method leaves little to be desired: The cover-glass is held in the forceps by one corner, film up, a large drop of carbolic fuchsin is placed on it so as to entirely cover the upper side, and it is heated directly over the flame until it steams fully or even boils, care being taken not to let any part become dry. If necessary, a second drop of the staining fluid is added as the first evaporates, and the heating is continued for, perhaps, half a minute. After rinsing most of the staining fluid off under the faucet, the cover is dropped into *very* dilute nitric acid until the red color changes to a dull olive, or begins to disappear entirely (which usually requires very few seconds), after which it is at once moved about in a dish containing a considerable quantity of alcohol until the dye, which has resumed its original red color, ceases to be given off in clouds. If the film still retains any noticeable color, it is dipped into the acid once more and again washed in alcohol, after which it is counter stained with methylene blue, rinsed in water, and may be at once examined, or dried and mounted in balsam. By this method, a good permanent preparation may be had under the microscope within five minutes of the time when the needle is first dipped into the sputum for the transfer of a little to the cover-glass.

While there are some advantages in floating cover-glass preparations upon the staining fluid, as the author recommends, it is more common to keep dilute alcoholic solutions (No. 4) of methylene blue, gentian violet, and fuchsin, as well as the standard alkaline blue (No. 10) and carbolic fuchsin (No. 7) in 2 oz. wide-mouthed bottles closed with bored corks through which dropping-tubes pass into the middle of the bottle, so that a drop of the desired staining-fluid may easily be allowed to fall upon the hardened film, heat being applied when necessary by holding the cover above a Bunsen burner until steam begins to rise.—W. T.]

Staining Spores on the Cover-glass.—When bacilli which contain spores are stained by the above methods, the spores remain uncolored, so that they appear as colorless spots within the strongly colored rods. This indisposition of the spores to receive coloring matters can be overcome in various ways, and it was simultaneously shown by Buchner and Hueppe that a prolonged exposure to heat, which even unfits the vegetative cells of bacteria for staining (*cf.* p. 144, Ehrlich), exactly adapts the spores to the reception of basic aniline colors in aqueous or dilute alcoholic solutions.

(VI.) If, instead of passing the cover-glass through the flame three times to harden it (p. 144), it is flamed ten times, or heated fifteen to twenty minutes at 120° to 180° C., and then stained with an aqueous solution of one of the usual basic colors, the spores become deeply stained; but the staining of the spores is isolated, because, as has been said, the rods have lost their receptivity for the dye.

There is, however, a means of obtaining double-stained preparations, in which the spores take one color, the rods, another: namely, by an extremely intense application of the Ehrlich tubercle method.

(VII.) The cover-glass film is hardened in the flame as usual (three times), stained for an hour floating on hot aniline fuchsin (p. 144), rinsed in water, decolorized in 25 parts hydrochloric acid and 75 of alcohol, and counter-stained in a saturated aqueous solution of methylene blue.

There is a great difference in the readiness with which the spores of different bacilli may be stained. [*B. Megatherium* and *B. subtilis* are good species to practise with, *B. anthracis* is among the more difficult. The endospores of yeasts may be stained by the same methods, decolorization being here effected by alcohol without the addition of an acid. Ziehl's carbolic fuchsin, heated upon the cover, as in tubercle staining, gives excellent results, but it must be renewed several times and the boiling correspondingly prolonged.—W. T.]

Staining the Flagella of Bacteria.

In the case of certain large bacteria, *e.g.*, the forms of *Beggiatoa roseo-persinica* first thoroughly studied by Warming, —the flagella can be seen by aid of a good objective without any preparation whatever. In the case of smaller forms, they can only be demonstrated by staining them, or even by means of staining and photography; but none of the methods so far indicated are applicable to this purpose. Koch, however, succeeded in demonstrating them by the use of a saturated aqueous solution of extract of logwood, added to the fluid containing the bacteria. His permanent preparations were made as follows:

(VIII.) The bacteria are dried on the cover-glass, stained with an aqueous solution of logwood, laid in dilute chromic acid (which forms a dark brown combination with the dye), dried, and mounted in balsam.

Neuhaus advises the replacement of chromic acid by neutral sodium bichromate, prepared by adding 5-per-cent soda solution drop by drop to dilute chromic acid. He recommends the following as the most certain means of staining the flagella:

(IX.) The cover-glass preparation is boiled for five minutes upon black logwood ink ("Kaisertinte"), from which it is placed for fifteen minutes in dilute neutral bichromate of sodium, this being repeated several times.

Staining Bacteria in Sections.

When it is only necessary to demonstrate the presence of bacteria in the various organs, it often suffices to make cover-glass preparations by rubbing on it a little of the juices from a fresh cut surface; but a more exact examination of the number of bacteria, and of their distribution in the tissues, can be made only by means of sections. Before Weigert provided methods of staining bacteria *in situ*, no means of demonstrating them existed, except by the use of acids and alkalies, to which they show great resistance.

(X.) When a fresh section, or one hardened in alcohol, is treated with strong acetic acid or dilute (2 per cent) solution of caustic potash or soda, it becomes almost transparent, only the bacteria resisting the acid or alkali, and so becoming evident, especially when collected in masses in or without the vessels; hence such nests of cocci were seen and described long before their nature was known, *e.g.*, by Beckmann, in the vessels of the kidney. Later, the potash method came to play an important part in the study of pyæmia early in the seventies, *e.g.*, in the work of Hj. Heiberg; and quite recently it has been used in isolated cases. It is thus used to render collections of typhoid bacilli evident in the tissues; and, even before knowing of Koch's investigations Baumgarten had seen tubercle bacilli by the use of the potash method. Still, in general, this has lost its importance since the introduction of staining methods.

One condition of a good staining of sections is their being well and completely hardened. The organs are cut into small pieces and hardened in a large quantity of absolute alcohol.

[If the bottle in which the pieces are to be hardened is filled half full of loose cotton, so that the tissues are kept near the top of the alcohol, the portion of this which becomes more charged with water tends to sink to the bottom because of its greater specific gravity, and it has been claimed that in such cases more thorough hardening is effected.—W. T.]

As thin sections as possible are cut from the well-hardened tissues by the use of a razor or microtome.

[In Koch's laboratory it is now customary to use a drop of melted glycerin-jelly for attaching the organ to a cork, by means of which it clamped in a Schauze or other sledge-microtome, without any sort of imbedding. If perfectly hardened material is used, fairly good sections are obtained in this way, the sectioning of course being done under alcohol; and there is no danger of the proper staining of the bacteria being interfered with. But in many cases, where the organs to be sectioned are rather large, brittle, etc., and the bacteria will not be injured by the preparatory treatment, the tissue may be imbedded in celloidin for sectioning in the usual way under alcohol; or it may be impregnated with paraffin melting at about 50° C., and imbedded in this for sectioning dry—preferably with some form of rocking-microtome. In the latter case, if the organ is small, ribbon sections may be obtained and stained on the slide, in the way so generally employed now by histologists, and especially embryologists; or the paraffin may be removed by placing the sections in turpentine for a few moments, and afterward in alcohol, after which they are stained individually as if cut without imbedding.—W. T.]

The staining of sections is essentially the same as staining cover-glass preparations; but it must be observed that as a rule they require a longer treatment with the staining-fluid, are less tolerant of heating in the latter, and require more powerful decolorizing media for differentiating the bacteria, while anhydration before they can be mounted in balsam is rarely effected by drying, but by the use of alcohol or anilin oil (Weigert). To prevent these fluids from at the same time partly decolorizing the preparation, Kuehne has recently adopted the plan of anhydrating with a fluid tinged with the same dye used in staining the bacteria (No. XIII.).

Kuehne advises spreading differentiated sections of glanders material upon the cover-glass, blowing them dry (*cf.* p. 146), laying the cover-glass (with the section upward) upon a glass plate, where it is slightly warmed, not to exceed 30° C., over a spirit lamp, until the section becomes transparent, when, after lying on the warm plate for five minutes, it is cleared with ethereal oil and passed through xylol into balsam [so that the use of alcohol is entirely avoided because of its marked decolorizing action on the bacteria].

For the microscopic examination of sections which contain stained bacteria, a sub-stage condenser is used. The Abbé condenser is the most perfect of the various models in use, but the cheaper Dujardin and similar condensers, which are readily used on smaller stands in place of the well-diaphragm, are entirely satisfactory.

Koch first called attention to the importance of the condenser in bacteriological microscopy. By means of this instrument, as he expresses it, it is possible to efface the "structure image" and so to get rid of its disturbing and concealing influence on the "color image." By the former name he designates the image of lines and shadows which reveals to us the structure of the tissue, and which results from diffraction of the rays of light during their passage through the preparation, different parts of which (nuclei, fibres, etc.) differ in refractive power from the substratum in which they lie By the "color image," he refers to the image of colored parts, which, if very small, may be completely concealed by the lines and shadows of the structure image. By the use of an Abbé or similar condenser [with full opening], the preparation is illuminated with so broad a cone of light that the tissue appears as a structureless plane, on which the colored parts, large and small, stand out sharply. It is easy to convince one's self of the great advantage of using such a condenser, for a section which is full of stained bacilli of mouse septicæmia, may appear free from bacteria when examined with ordinary illumination with the concave mirror and diaphragm, while the condenser resolves a mass of colored rods within it.

What is here said of the examination of stained bacteria in sections, is also true of cover-glass preparations, in which the bacteria frequently occur upon or within elements the structure images of which mask them. On the other hand, it must be remembered that the examination of unstained bacteria should always be made without a condenser; with the finest diaphragm which gives sufficient illumination; or, what amounts to the same thing, with the condenser lowered as far as possible.

Methods of Staining Sections.

Several of the numbered methods which follow have already been described for cover-glass preparations, but are repeated here with the changes necessary for sections.

(XI.) *Weigert's Original Method* (1876).—The section is laid for a rather long time in a quite strong solution of methyl violet, washed out in dilute acetic acid, anhydrated in alcohol, cleared in oil of cloves, and mounted in balsam.

Weigert's improved method (1881) employs an aqueous 1-per-cent solution of gentian violet, BR., from which the section goes to absolute alcohol, clove oil, and balsam. (Weigert recommended gentian violet, BR., especially because it is not so easily removed from the bacteria when the section is treated with alcohol and clove oil.)

(XII.) *Loeffler's Potash Blue Method.*—The sections are placed for a few minutes in alkaline methylene blue (No. 10), differentiated for some seconds in 0.5-per-cent acetic acid, anhydrated in alcohol, cleared with cedar-oil, and mounted in balsam.

(XIII.) *Kuehne Universal Method.*[8]—The section is laid in carbolic methylene blue (No. 6), as a general thing for half an hour (but for leprosy, two hours), rinsed with distilled water, treated with dilute hydrochloric acid (No. 12) until the color is pale blue, rinsed in lithium-water (No. 14), laid in distilled water for some minutes, dipped for anhydration into absolute alcohol (which may be colored with methylene blue, p. 590), before it goes into a salt-cellar of aniline-oil colored with methylene blue, where in a few moments it becomes anhydrated without being decolorized, then rinsed in pure aniline oil, cleared in turpentine, and when all aniline oil has been carefully removed by treatment with xylol, usually renewed once, it is mounted in balsam. When the aniline oil is not completely removed, the balsam becomes brown with time.

The differentiation in hydrochloric acid is the most difficult point in the process. Thin sections need only be dipped into it; and in any case a sharp watch must be kept upon the degree to which decolorization has progressed, so that the section may be rinsed in lithium water at once when the right shade is reached.

(XIV.) *Koch's Isolated Staining Method.*—After staining in an aqueous solution of fuchsin, methyl violet, or methylene blue, the sections are washed out in a half saturated solution of carbonate of potassium (No. 15 *a*), then passed through alcohol, xylol, and balsam. Weigert was the first to undertake an isolated staining of bacteria in tissues, since by the use of carmine, followed by washing in glycerin acidulated with hydrochloric acid, he stained clusters of micrococci in 1871.

Malassez and Vignal, after staining in methylene blue, wash in No. 15 *b*, etc.

(XV.) *The Gram Method.*—The sections are transferred directly from absolute alcohol into aniline gentian violet, and treated like cover-glass preparations (No. III.), except that they are anhydrated in absolute alcohol and cleared in cedar-oil preparatory to being mounted in balsam. For Weigert's modification of the method, see No. XIX., *infra.*

When sections stained by this method are also treated with a nuclear stain like carmine, a very pretty double staining is obtained. Fraenkel recommends the following method, which is also thought to more frequently escape the troublesome granular precipitates which now and then appear in the preparation when the original Gram method is used.

(XVI.) The sections are transferred from alcohol to strong picro-carmine, where they remain half an hour, are rinsed in 50-per-cent alcohol, then go for half an hour into unsaturated aniline gentian violet—prepared by dropping a few drops of the saturated alcoholic solution (No. 1) into a watch-glass containing aniline water (No. 5) until the mixture begins to be opaque—from which they are placed directly in the iodine water for three minutes, then into alcohol, where the red color reappears and they become anhydrated, when they go through cedar-oil into balsam.

(XVII.) *Koch-Ehrlich-Weigert's Tubercle Method.*—The sections are laid in aniline gentian violet (No. 9) for twelve hours, moved about for a few seconds in 25-per-cent nitric acid, rinsed in 60-per-cent alcohol until they show only a slight blue color, counter-stained for some minutes in a saturated aqueous solution of vesuvin, rinsed in 60-per-cent alcohol, anhydrated in absolute alcohol, cleared with cedar-oil, and mounted in xylol balsam.

The tubercle bacilli are also stained by Kuehne's method

(XIII.) and the Gram method, but where a differential diagnosis is required decolorization must be effected by strong mineral acids (*e.g.*, as in Nos. XVII. and XVIII.); for it is its resistance to decolorization by mineral acids which distinguishes the tubercle bacillus from all other known species except the bacillus of leprosy, which, however, differs from it in staining readily with a simple aqueous solution of methyl violet or fuchsin. The so-called smegma-bacilli also show a decided power to resist the action of mineral acids, but they are readily decolorized by subsequent treatment with alcohol, which is not the case with either the tubercle or leprosy bacillus. The smegma-bacilli are probably only different species of putrefactive bacteria which, from growing in the smegma, have become so impregnated with oily matters that they have become difficult to stain. This view is supported by the fact that it has proved possible to change the behavior of other species toward staining methods, by growing them in substances containing butter, etc. (Bienstock). On the other hand, the diagnosis of smegma-bacilli may be attended with some difficulty as compared with Lustgarten's syphilis bacillus, the etiological relation of which to the disease is still very problematical, although it has been demonstrated in syphilitic secretions and neoplastic growths. Like the syphilis bacillus, after being stained on the cover-glass for twenty-four hours in aniline gentian violet, they remain colored after treatment for ten seconds with 1.5-per-cent solution of potassium permanganate, followed by sufficient washing in sulphurous-acid water to entirely decolorize the preparation. The syphilis bacillus cannot be confounded with that of tubercle, because it is very easily decolorized by mineral acids.

(XVIII.) *Ziehl-Neelsen Tubercle Method.*—Sections are more rapidly stained by laying them for fifteen minutes in carbolic fuchsin (No. 7), agitating them a few seconds in 25-per-cent sulphuric or nitric acid, rinsing in 60-per-cent alcohol until only a light rosy color remains, counter-staining for some minutes in a saturated aqueous solution of methylene blue, and rinsing, anhydrating, and mounting as before.

For staining bacteria in sections of hardened gelatin-cultures, see Neisser, "Centralbl. f. Bakteriologie," 1888, III., 506.

Moulds.

The following notes on mounting moulds refer chiefly to the directions of O. Johan-Olsen.

Temporary preparations are obtained by laying the mould for a few minutes in dilute alcohol, then in weak ammonia, which is removed with filter-paper, the surplus washed out with water, and this replaced by glycerin, in which the examination is made.

Permanent preparations are made in the same manner, after staining the material in osmic acid, which is kept for use in a 0.5-per-cent solution in a dropping-bottle that has been well cleansed with alcohol and ether and is stopped with glass. Two drops of this are added to eight drops of water in a watch-glass (*i.e.*, about 0.1 per cent), in which the mould is allowed to lie for an entire day; or, if the 0.5-per-cent solution is used directly, a few minutes suffice. The preparation is then washed with alcohol, followed by distilled water. If it is still too black, it is decolorized sufficiently in weak ammonia, followed by water, and mounted in glycerin. Preparations stained with osmic acid may also be further stained by saffranin, usually a very dilute solution, in which they are allowed to remain for a long time.

Preparations of an entire culture can be made upon the slide. Usually the cultures are obtained from fluid media (*e.g.*, Brefeld's slide-cultures, *supra*, p. 100), and those which are not too far advanced are chosen. The colony being placed on a slide, a cover-glass is let down upon it carefully—especially in case of the erect forms—and it is allowed to dry down so that the cover-glass does not move around (usually only a few minutes being needed), when 0.1-per-cent osmic acid is added at one side and drawn under by applying bibulous paper to the other edge of the cover. After a proper time the acid is thoroughly washed out with water that is drawn under the cover in the same way, and, either with or without saffranin staining, mounted by drawing glycerin under and sealing in the usual way; or else dilute alcohol is used instead of the osmic acid, and followed by ammonia, etc., as indicated above for temporary preparations.

For demonstrating moulds in sections, Hueppe recommends

the methylene blue universal method (Nos. XII. and XIII.). Ribbert advises the following modification of the Gram method:

(XIX.) *Weigert's Fibrin Stain.*—The section hardened in alcohol is stained for a long time in saturated aniline gentian violet (No. 9), rinsed in 0.7-per-cent solution of table salt, and transferred to the slide, where the other steps are taken. The superfluous water is removed with filter-paper, iodine water (No. 16) is dropped upon it and removed with filter paper, and a couple of drops of aniline oil allowed to fall upon the section for the purpose of at once decolorizing and anhydrating it.

As the aniline oil soon becomes dark, it must be removed and replaced by fresh oil once or twice, after which it is entirely replaced by xylol and the section mounted in balsam.

Actinomyces.

When it is only desired to obtain a sure diagnosis of *Actinomyces* cushions in pus or sections, treatment with acetic acid or alkalies and examination in glycerin suffices. For staining cover-glass preparations, Fraenkel employs the Gram method, letting the section stay in aniline gentian violet for twenty-four hours, and in the iodine solution for fifteen minutes. The Weigert-Gram method (No. XIX.) appears suited to staining sections for *Actinomyces.*

Amœboids.

The amœbæ or amœboid organism detected by Laveran and subsequently more fully studied by Marchiafava and Celli, Golgi, Osler, and others, and held to be the cause of malaria, are not adapted to staining by any of the above methods. It is, indeed, possible, when cover-glass preparations of the blood are treated with methylene blue or vesuvin, to find within the red corpuscles larger or smaller irregular spots, colored blue or brown (Marchiafava and Celli). These are the stained amœbæ; but a clear view of all the developmental forms of the parasite is obtained only by examination of the fresh blood. Laveran gives the following directions for this: The amœbæ are most easily found in patients already anæmic, and who have not taken quinine, at any rate just before the examination. The blood is obtained just before, or in the beginning of an attack, by a needle puncture in the finger-tip or lobe of the ear, after careful cleansing with water and alcohol. The surface of the drop which exudes is touched with a well-cleaned slide, and the small drop of blood which adheres is quickly covered with a cover-glass, and sealed with paraffin when the layer of blood is thin enough for examination. An enlargement of 400 to 500 diameters is sufficient for the investigation, places being chosen where the red corpuscles lie flat and isolated. Few parasites are usually to be seen, rarely as many as eight to fifteen in the same field, so that one must be prepared to spend some time in searching for them. The pigment granules that occur in most of the amœbæ give a clue in the search. Richard recommends the addition of a drop of dilute acetic acid for the rapid demonstration of these parasites, as it removes the annoying red blood-corpuscles without destroying the parasites. According to Laveran, the addition of water is better, as it does not kill the amœbæ so soon as acetic acid does, but allows them to continue their peculiar movements.

BIBLIOGRAPHY.

1. Fortschritt. d. Medicin, 1886, iv.—2. Meddelelser f. Carlsberg Lab., ii., 218.—3. *Cf.* Fortschr. d. Med., 1888, No. 1.—4. Zeitschr. f. Hygiene, 1887, 521.—5. Ann. de Micrographie, i., 153.—6. Sur la culture des microbes anaërobies (Annales de l'Institut Pasteur, 1887, No. 2).—7. Beiträge zur Kenntniss des Sauerstoffbedürfnisses der Bacterien (Zeitschr. für Hygiene, 1886, i., 115).—8. From Kuehne: Praktische Anleitung zum mikroskopischen Nachweis der Bakterien im thierischen Gewebe, Leipzig.

INDEX.

www.ingramcontent.com/pod-product-compliance
Lightning Source LLC
Chambersburg PA
CBHW030242170426
43202CB00009B/600
* 9 7 8 3 7 4 3 6 8 6 9 1 5 *